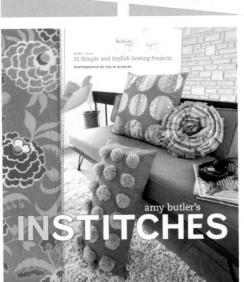

Craft:™

Volume 01

Features

30

This beaded *Backyard* by Liza Lou is comprised of more than a million beads.

46

Columns

40

Crafter Profiles

Inside the lives and workshops of:

ON THE COVER

Stitched zombies, cute creatures, and robots are the most popular items on Etsy. The most popular seller on Etsy is Beth Doherty, aka Gourmet Amigurumi, who's influenced by Japanese artisans. She shows us how to crochet our own robot on page 123. Cover doll by Jess Hutch and cover photography by Dwight Eschliman.

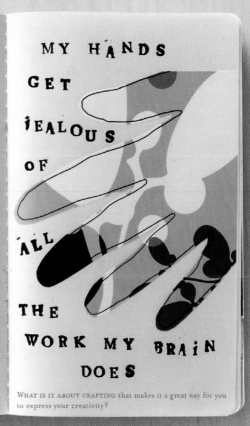

MY HANDS GET JEALOUS OF ALL THE WORK MY BRAIN DOES

WHAT IS IT ABOUT CRAFTING that makes it a great way for you to express your creativity?

ELISABETH ELDE / MINNEAPOLIS, MN

WHAT DO YOU HAVE TO SAY ABOUT YOUR CRAFT?

WHAT DOES YOUR CRAFT SAY ABOUT YOU?

SHARE YOUR STORIES AT FISKARS.COM

FISKARS®

Craft:™ Projects

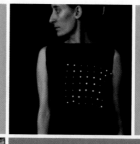

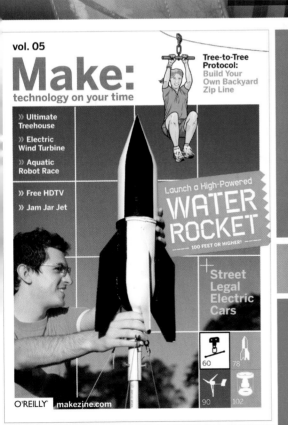

Volume 01

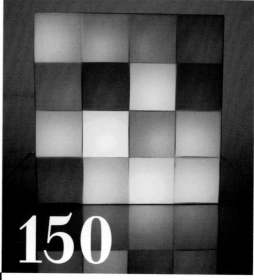

150

20

Make Cool Stuff

106

The Crafting of Craft:

Welcome to the new magazine for the new craft movement.

Originally, **CRAFT magazine was** supposed to be a one-off craft-themed issue of MAKE magazine, featuring some of the cool but "craftier" DIY projects that enthusiastic readers sent in.

For instance, Ph.D. student Leah Buechley from University of Colorado at Boulder sent us photos of her programmable LED tank top that flashes the Game of Life. This project definitely has the elements of a MAKE project — it involves soldering, LED technology, and programming. But there are also craft elements that don't quite jibe with MAKE's harder-edged sensibility: it requires a sewing machine, sewing skills, fabric, and a pattern. And unlike the projects in MAKE, where the end result is more about function than form, it's essential for this project to be as aesthetically attractive as it is useful.

We also received a bunch of simpler projects that had MAKE's DIY tech flavor but, again, also involved "craftier" skills, materials, and an emphasis on the visual outcome. Some of these include hacking action figures, knitting steel medieval armor, making notebooks out of floppy disks, and building lamps out of vintage kitchenware.

While we were putting our CRAFT issue together, we became aware of the exploding craft movement taking place all around us. This DIY renaissance embraces crafts while pushing them beyond traditional boundaries, either through technology, irony, irreverence, and creative recycling, or by using innovative materials and processes.

Crafters are creating websites like Craftster, Get Crafty, Crafty Chica, The Anti-Craft, The Church of Craft, Sew Darn Cute, Hobby Princess, Extreme Craft, The Craft Mafias, The Switchboards, SuperNaturale, and KnitKnit to build community, exchange ideas, and sell their projects. They're also networking at real-life craft/DIY fairs like Bazaar Bizarre, Renegade Craft Fair, Stitch, Swap-O-Rama-Rama, and Felt Club.

When MAKE had its first Maker Faire in San Mateo, Calif., last April, we were blown away by the attention our special craft areas received. An excited crowd swarmed our craft demonstrations (like how to use a Gocco printing machine), our tech-geek-meets-haute-couture fashion show (featuring Diana Eng's amazing clothing), and craft booths that offered handmade treasures such as reed switch necklaces, skull-patterned baby clothes, one-eyed monster dolls, espresso-charged soap, and scads of other items.

It was during the Faire that we decided to turn CRAFT into its own quarterly magazine — a sister publication to MAKE. This first issue of CRAFT has over 20 projects, including how to make an LED shirt, how to embroider a skateboard, how to convert a dud pair of shoes into awesome knitted boots, and how to make a felted iPod cocoon.

Beyond projects, CRAFT is also filled with features on topics such as crocheting math equations and the fascination with knitted robots, zombies, and other creepy-cute creatures.

The new craft movement encourages people to make things themselves rather than buy what thousands of others already own. It provides new venues for crafters to show and sell their wares, and it offers original, unusual, alternative, and better-made goods to consumers who choose not to fall in step with mainstream commerce. Crafting empowers people by allowing them to create something useful. If you need something, just make it yourself. And make sure you check out craftzine.com for more crafty resources. ✄

Craft: ™
transforming traditional crafts

EDITOR AND PUBLISHER
Dale Dougherty
dale@oreilly.com

EDITOR-IN-CHIEF
Carla Sinclair
carla@craftzine.com

CREATIVE DIRECTOR
David Albertson
david@albertsondesign.com

MANAGING EDITOR
Shawn Connally
shawn@craftzine.com

ART DIRECTOR
Sara Huston

ASSOCIATE MANAGING EDITOR
Goli Mohammadi
goli@craftzine.com

DESIGNERS
Gerry Arrington
Sarah Hart
Kirk von Rohr

ASSOCIATE EDITOR
Natalie Zee
nat@craftzine.com

ONLINE MANAGER
Terrie Miller

CONTRIBUTING EDITOR
Phillip Torrone

ASSOCIATE PUBLISHER
Dan Woods
dan@oreilly.com

STAFF EDITOR
Arwen O'Reilly

CIRCULATION DIRECTOR
Heather Harmon

COPY EDITORS/RESEARCH
Colleen Gorman
Keith Hammond
Matt Hutchinson
Sanders Kleinfeld
Rachel Monaghan
Laurel Ruma
Marlowe Shaeffer

ADVERTISING COORDINATOR
Jessica Boyd

SALES & MARKETING ASSOCIATE
Katie Dougherty

MARKETING & EVENTS COORDINATOR
Rob Bullington

CRAFT TECHNICAL ADVISORY BOARD:
**Jill Bliss, Jenny Hart, Garth Johnson,
Leah Kramer, Alison Lewis, Matt Maranian,
Ulla-Maaria Mutanen, Kathreen Ricketson**

PUBLISHED BY O'REILLY MEDIA, INC.
Tim O'Reilly, CEO
Laura Baldwin, COO

Visit us online at craftzine.com
Comments may be sent to editor@craftzine.com

For advertising and sponsorship inquiries, contact:
Dan Woods, 707-827-7068, dan@oreilly.com

Customer Service cs@readerservices.craftzine.com

Contributing Artists:
Meiko Arquillos, Vincent Atos, Kenny Braun, Tim Brown, Frick Byers, Howard Cao, Daniel Carter, Hank Drury, James Duncan Davidson, Dwight Eschliman, Philip Heying, Jemma Hostetler, Tim Lillis, Christopher Lucas, Steve Mack, Jason Madera, Dave McMahon, Anitra Menning, Quinn Norton, Brian Sawyer, Susan Sheridan, Jen Siska, Peter Wrenn, Dominik Enerek Zacharski

Contributing Writers:
Aram Bartholl, Gareth Branwyn, Mary Belton, Jill Bliss, Susie Bright, Leah Buechley, Reagan Copeland, Beth Doherty, Emily Drury, Mark Frauenfelder, Holly Gates, Jenny Hart, Mister Jalopy, Xeni Jardin, Leah Kramer, Steve Lodefink, Matt Maranian, Tina Marrin, Kathy Cano Murillo, Ulla-Maaria Mutanen, Bob Parks, Kristina Pinto, Jean Railla, Kirk von Rohr, Michael Shapiro, Julia Szabo, Annalee Newitz, Gretchen Walker

Interns: Matthew Dalton (engr.), Adrienne Foreman (web), Jake McKenzie (engr.), Ty Nowotny (engr.)

 CRAFT is printed on recycled paper with 10% post-consumer waste and is acid-free. Subscriber copies of CRAFT, Volume 01 were shipped in recyclable plastic bags.

Vol. 01, October 2006. CRAFT (ISSN 1932-9121) is published 4 times a year by O'Reilly Media, Inc. in the months of January, April, August, and October. O'Reilly Media is located at 1005 Gravenstein Hwy. North, Sebastopol, CA 95472, (707) 827-7000. SUBSCRIPTIONS: Send all subscription requests to CRAFT, P.O. Box 17046, North Hollywood, CA 91615-9588 or subscribe online at craftzine.com/subscribe or via phone at (866) 368-5652 (U.S. and Canada), all other countries call (818) 487-2037. Subscriptions are available for $34.95 for 1 year (4 issues) in the United States; in Canada: $39.95 USD; all other countries: $49.95 USD. Application to Mail at Periodicals Postage Rates is Pending at Sebastopol, CA, and at additional mailing offices. POSTMASTER: Send address changes to CRAFT, P.O. Box 17046, North Hollywood, CA 91615-9588.

Contributors

Jen Siska (*101: Silk-screening* photography) is a self-proclaimed quirky girl who loves to photograph people. Typically photographing her subjects in ordinary settings using only natural light, each picture is a celebration of the overlooked things in life. A San Francisco resident, Jen's favorite food is birthday cake. She recently ran her first solo art show (which even included handmade frames for the photos) for an entire month at Rowan Morrison Gallery in Oakland, Calif. Collections of Jen's eye candy reside at jensiska.com.

A diehard Chicagoan, **Beth Doherty** (*Cro-bot*) is the crocheting mastermind behind gourmetamigurumi.com, home of the tastiest treats around. Even though she was hospitalized for amnesia last year, she still can't stop dreaming about yarn. Known for her unique style and loving attention to detail, Beth's ridiculously adorable creations take forever to craft and no time at all to sell. Her tribe consists of her husband and two cats. Keep an eye out for Beth's new amigurumi book coming out this fall from Lark Publications.

Matt Maranian (*Ant Farm Room Divider* and *Raw But Refined*) is a designer and best-selling author of the DIY/style books *PAD* and *PAD Parties*, as well as the celebrated — albeit infamous — guide to Los Angeles, *L.A. Bizarro*. A half-time retailer, he and his wife Loretta own Boomerang, a new and vintage clothing store in Brattleboro, Vt., and live just outside of town nestled among the birch trees in their modern "vacation cabin," originally built for *Woman's Day* magazine as a model home in 1963. A sometime drummer and full-time bon vivant, he's currently at work on a new book.

Leah Buechley (*The Electric Tank Top*) has been obsessed with art, craft, mathematics, and engineering for as long as she can remember. Her research interests include electronic textiles and wearable computing, tangible interfaces, education, and human computer interactions. She is delighted to have found a place where she can do everything she loves: the Craft Technology Group at the University of Colorado (www.cs.colorado.edu/~buechley). Isn't it wonderful that you can get a Ph.D. building light-up clothes and jewelry?

Leah Kramer (*Customized Kitsch*) has been crafting ever since she could hold a pair of safety scissors. Somewhere along the line she thinks she inhaled too much glue because now she is attracted to crafts that are irreverent, ironic, kitschy, or cleverly eco-friendly. In 2003 she started craftster.org — an online community for rebel DIYers. She resides in Boston where she is part-owner of a handmade crafts boutique called Magpie and is an organizer of the Bazaar Bizarre craft fair. Leah is also the author of *The Craftster Guide to Nifty, Thrifty, and Kitschy Crafts*.

Fueled by loose glitter and caffeine, **Kathy Cano Murillo** (*Fiesta Explosion Flower Pots*) lives in Phoenix with her husband, two kids, three chihuahuas, and her favorite tool, a chopstick. Though coffee, chocolate, and chili are her three favorite flavors, she's tasted (accidentally) glitter, varnish, paint water, and kindergarden craft paste (not by accident). Check out Kathy's crafty projects and mucho colores offerings at craftychica.com.

Jean Railla
Modern Crafting

>> Jean Railla is the founder and editor of getcrafty.com and the author of *Get Crafty: Hip Home Ec* (Broadway Books). She lives in Greenwich Village with her husband, two rapscallion sons, and a gazillion half-finished craft projects. jean@getcrafty.com

Why Making Stuff Is Fashionable Again

Is it just me, or did crafting become ubiquitous overnight? In 1998, when I started Get Crafty (getcrafty.com), knitting, sewing, and "keeping house" seemed quaint, ironic even, but hardly a nationwide trend. Then it happened. The Style Channel launched the hilarious *Craft Corner Death Match*; *Time* devoted several pages to DIY fashion in their article "Pretty Crafty"; Debbie Stoller's *Stitch-n-Bitch* hit *The New York Times'* bestseller list; and publishing houses released a plethora of cool craft books (including my own), with kits, TV shows, and other synergistic multimedia opportunities.

So why now? Why after feminism, the Industrial Revolution, and the pervasiveness of the Gap are the young and the beautiful suddenly knitting baby blankets, hand-cranking mango-mint ice cream, and sewing vintage fabric skirts?

I have a few theories.

Theory Number 1: Painters, photographers, rock stars, actors, and designers are the folks our culture holds up as heroes. While not all of us can earn a paycheck from our artwork, crafting allows us to make art out of everyday life. Choosing the texture and color of a yarn, creating a pattern, knitting it, wearing it — it's a full creative process. Everyone from twenty-somethings to baby boomers were raised to believe that making art is the ultimate contribution — and the most gratifying. Crafting allows us to embody this bohemian ideal while paying the rent.

Theory Number 2: Feminism was successful. The leaders of the Women's Movement of the 1960s and 70s rejected the domestic as a symbol of their oppression, but they unwittingly paved the way for all those ironic crocheted sushi rolls that kids love nowadays. Think of the guitarist with the handmade skull and crossbones wrist guard. Even a few years ago, he would've been ridiculed for knitting. Now he's a craftster. Or the fashion editor who brags about her crocheted skirt. With a decent paycheck, her own apartment, and an outlandish social life, other women can't scorn her happy hooking or consider her a "granny."

By leveling the playing field between men and women (at least in the bottom rungs of the work-force), feminism opened the door for all of us to value typically feminine art forms.

Theory Number 3: We work at computers all day. Crafting allows us the experience of the tactile world, the non-virtual, the "real." In a world where only a few actually manufacture products, making something that you can touch, wear, or inhabit is satisfying on an almost spiritual level. Let's just say it feels good.

Theory Number 4: Crafting is a political statement. With globalism, factory labor, and sweatshops as growing concerns, and giant chains like Starbucks, McDonald's, and Old Navy turning America into one big mini-mall, crafting becomes a protest. By MIY (Making It Yourself), we vote with our wallets and assert our individuality, knowing that no one will have the same hand-knit sweater or silk-screened T-shirt.

Given all this, it's hard to bemoan the popularization of crafting. What is "selling-out" if you only encourage creativity? Although books, kits, and TV shows can inspire you, at the end of the day, it's just you and your craft supplies.

The point of crafting is to be in touch with one of the things that make us human — our ability to make stuff. And if this spreads like organic honey on a hot stove, then I'm all for it. ✖

ETSY + CRAFT CONTEST WINNERS

About 300 Etsy crafters entered our "incorporate the **Craft:** logo into your craft" contest to introduce the magazine to the Etsy community. Some were fun, some were quirky, all were amazing in their detail and thoughtfulness. Here are the editors' top choices. Check out their comments and more information about the artists at craftzine.com/go/etsy.

» Interactive Pendant BY KATHRYN RIECHERT

Kathryn attended the Savannah College of Art and Design. She quickly developed a love for designing and making jewelry. Kathryn believes jewelry can become a very personal thing, with tremendous sentimental value, passed through families for countless years. The ultimate compliment is seeing someone wearing her jewelry because a part of her goes into every piece she makes. kathrynriechert.com

« Outta This World Cigar Box Purse
BY EMOTIONAL BAGGAGE

As long as she can remember, Tamara Rosas has had a paint brush or some form of glue or sewing needle in her hands. Encouraged by her grandma, she likes doing all forms of craft and has recently fallen in love with sewing all over again. In fact, it's safe to say that sewing is her one true craft love … for now. emotionalbaggage.etsy.com

» Crafts DO Grow on Trees BY FELLNWARD

Julia Ward and Cara Fellrath met in July 2003 and became best friends at first sight. Both are 24-year-old, classically trained artists; Julia's concentration is in painting, Cara's is in sculpture. They spend all their free time making stuff: scarves, T-shirts, bags, anything they can from recycled materials. fellnward.etsy.com

« Betty the Retro Robot Waitress BY ARTSY

Cat Bishop's work explores the disconnect between the self each of us harbors inside and the self we present to the world. Her work is about the cobbling together of a self more presentable than the one we know. While the work is, in a deep way, about self-scrutiny and self-flagellation, the artist clearly has fun with it. artsy.etsy.com

⚑ Geisha Tote
BY ENAMOR

Amie Miller's inspiration for crafting comes from her mother, who passed on her creativity and love of crafts. She's always liked to draw and in recent years has explored photography. She loves capturing things not normally seen with the naked eye. enamor.etsy.com

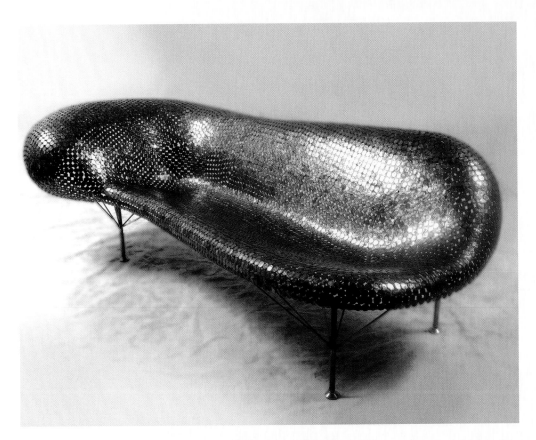

Sitting on Glass

When I first saw this chair made of 96 glass jars I immediately summed it up as aesthetics over function. It was beautiful, but no way could that glistening thing be comfortable. Then I sat in it and proved myself wrong. It's shocking how cozy so much glass can be.

As a Vermont steel-welder-turned-artist, **Johnny Swing** crafts found objects into practical works of art. He's known for his Nickel Couch (above), made with over 6,400 welded nickels, and his elegant chandeliers made of glass jars and satellite dishes.

He was inspired to create his Jar Chair when he noticed his son's baby food jars piled into garbage bags, waiting to be recycled. "They seemed too precious to discard. They were jewel-like."

It took Swing six weeks of intense design testing to come up with a formula for the chair, one that incorporates a hole pattern on aluminum sheets onto which he bolts the metal lids before screwing on the jars. "The final product looks design-engineered, but it wasn't. It was just tested over and over again."

He also switched from baby food jars — which he found too grueling to collect, sort, and clean — to Flint Sample Jars, which he could simply purchase in bulk.

When I asked Swing if he considers himself a crafter or an artist, he seems reluctant to directly answer the question. Instead he says, "What makes art is something that expands your intellect, and that's why the chairs are not strictly a craft. When you look at the Jar Chair, you're forced to ask questions, and in some ways be enlightened through the experience. Like, 'I'm going to sit on glass?' or 'How can this be comfortable?'"

With impeccable welding skills and an amazing eye for design, I'd say Swing straddles the worlds of both art and crafting quite exquisitely.

—Carla Sinclair

≫ **Obsessive Furniture:** johnnyswing/furniture.shtml

Photography by Jeff Baird

Not to Scale

Recycling is all well and good, but reusing is even better. Escama, a company that sells remarkably cute handbags made from soda can pull tabs and crocheted cord, may not use the whole can, but they make a good start. Crocheting with tabs is a traditional craft in Brazil, and Escama's unlikely founders (two American men, **Andy Krumholz** and **Eric Pedersen**, and a Brazilian woman, **Socorro Leal Schwiderski**) knew a good thing when they saw it. Working directly with the women in Brasilia's Cia do Lacre and 100 Dimensão cooperatives, the three friends adapted the traditional designs into what Escama calls a "fashion-forward accessory" (but what we call a really cool clutch).

Escama means "scale" in Portuguese — as in fish scale — but it could just as easily refer to scale of production. Though warned by apparel industry experts that mass-producing handmade items with co-ops was a disastrous idea, Escama has made over 3,000 bags in four different styles to date with impressive quality control, and they've expanded their workforce from 12 to 50 in the past year.

This doesn't mean they've lost sight of the individuals involved, however. Each bag comes with a tag inside printed with the name of the woman who made it, and you can look up each crochet artist's bio on their website.

"Promoting the cooperatives and the individual artist is fundamental to what we are about," says Krumholz. Next in the lineup are a few mystery products ("I won't let the cat out of the bag, but they're killer designs," Krumholz confides) as well as an online how-to for Escama's many crochet-addicted fans. Now *that's* what we call fair trade.

—*Arwen O'Reilly*

≫**Escama Bags:** escama.com

Photograph by Ryan Field

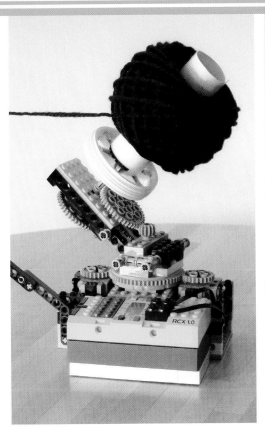

Wound Up with Legos

Photography by David North

Web developer **David North** didn't have a project in mind when he bought a Lego Mindstorms set and a bunch of other Lego Technics parts on eBay. He shelved them, allowing them to collect dust, while he waited for inspiration to hit him.

"I'd seen a lot of cool Lego projects online, like a machine that solved a Rubic's Cube and the Lego Segway — or Legway — but I thought it would be great to build a machine that actually performed a useful function," says North, who lives in York, England, with his girlfriend Mel Martindale.

Inspiration finally came after Martindale received many hanks of yarn for Christmas. "It was taking her ages to wind each one, and it's tricky to wind a ball by hand in such a way that it doesn't tangle in use." What she needed was a yarn winder, and he had just the toys to make it with.

North went online to see how yarn winders worked and based his machine on the most common type, "where the bobbin is angled at 45 degrees and works with a planetary type of motion. The yarn wraps around the bobbin at a 45-degree angle, going from the top of the ball to the bottom. The bobbin rotates slowly on its own axis to distribute each wrap of yarn around the ball evenly."

Using approximately 120 Lego pieces, two motors, four AA batteries, and a cardboard tube that fits around the bobbin to minimize friction, it took North approximately 12 hours — and a few versions of the machine — before he got it right. "Building the winder was a fun project and more of a challenge than I thought it would be, particularly getting it to the point where it was robust and wound attractive balls."

More than a year and a half later, the winder is still doing a fine job, and, despite the fact that a few gears have worn down and needed to be replaced and that it quickly sucks up batteries, Martindale never knits without it.

—Carla Sinclair

≫ **Lego Yarn Winder:** craftzine.com/go/yarnwinder

Knitting XXXL

Using two John Deere excavators donated by a local construction equipment dealer, artist **David Cole** and his assistants knit a huge American flag in July 2005 with two 20-foot knitting needles made out of telephone poles with sculpted points. Cole spent days up in a cherry picker, looping the stitches and directing the movement of the needles.

After casting off, *The Knitting Machine* was ceremoniously taken down, folded, and displayed in a glass case at contemporary art museum Mass MoCa in Western Massachusetts. "I didn't see it as a performance," Cole says about his time spent knitting up in the crane. "Knitting for me is a physical meditation on the creative process ... and the final object is an artifact of [that] process."

Cole began knitting in college to help him focus during lectures. He also took a sculpture class that gave him new insight into his knitting. Since then, he's become known for his knit artworks, which turn traditional ideas of knitting upside down.

Other recent pieces include teddy bears knit from lead, a baby blanket made from incredibly toxic porcelain fiber, and a Kevlar sweater. He's in the process of creating an installation of 25 giant teddy bears sewn from black roofing rubber, to be installed like a flotilla of buoys in a harbor. Notice a pattern? Dangerous materials mix with sweet subject matter in a dance likely to give you whiplash.

All of his work toys with the obsessive nature of craft, but his most successful pieces have many layers of meaning. While they don't yield their answers up easily, they aren't deliberately obscure either.

"I want my work to be conceptually accessible," he explains. "But if someone wants to engage critically and think about international policy and domination of space, great." And as for making a subversive gender statement? "Totally secondary. It's there, but that's not what it's about," says Cole flatly. "It's visual ballet."

—Arwen O'Reilly

>> **Big Knitting:** theknittingmachine.com

Photograph courtesy of Larry Smallwood and Dave Cole

BUSINESS REPLY MAIL

FIRST-CLASS MAIL   PERMIT NO 865   NORTH HOLLYWOOD CA

POSTAGE WILL BE PAID BY ADDRESSEE

O'REILLY®

Craft:

PO BOX 17046
NORTH HOLLYWOOD CA 91615-9588

BUSINESS REPLY MAIL

FIRST-CLASS MAIL   PERMIT NO 865   NORTH HOLLYWOOD CA

POSTAGE WILL BE PAID BY ADDRESSEE

O'REILLY®

Craft:

PO BOX 17046
NORTH HOLLYWOOD CA 91615-9588

Photograph by Jane Burns

Hero Hacker

As anyone who's ever perused the action figure aisles at Toys "R" Us or the meticulously appointed McFarlene Toys at the local comic shop can tell you, models of cartoon, movie, and comic book characters have become big business. But what if your comic love falls to those obscure second- or third-string characters never likely to see the inside of an injection mold? If you're **Bill Burns**, or a growing number of other "custom figure" makers, you go all recombinant plastic on your existing collection and hack together the heroes (or villains) that you long to see in 3D.

The recent explosion of the figure industry, especially among adult collectors, and the growth of the net alongside it, have led to a rapidly expanding community of figure hackers. Burns says: "For years, I looked at the 'Homemade Heroes' section in *Wizard* magazine [a comic enthusiasts' pub] and said to myself, 'I could do that.' After getting over the anxiety of cutting up perfectly good figures, I began to customize characters that I knew toy manufacturers would never produce."

Burns now has some 200 custom figures in his collection, including everything from a 70s-era Red Tornado (DC Universe) and the obscure Black Condor to dozens of DC and Marvel figures done in the style of the animated *Batman* and *Superman* TV shows. And then there's the beloved Jonny Quest, complete with his dog Bandit. While this remains a hobby for him, he does get commissions, sometimes from the cartoon/comic creators themselves. A few years back, he was hired by the Cartoon Network to create a Space Ghost figure for use on their website. Like most hobbyists, any money Burns makes he pours right back into his obsession.

For those who want to try their hand at figure hacking, Burns offers a few tips: "Start out simple, maybe just repainting a figure or sculpting a few small details, and then work your way up." An excellent how-to section and tons of figure hacks can be found at Casimir's Inanimate Objects (pilliod.net).

—*Gareth Branwyn*

≫ **Bill's Customs:** billscustoms.com

A Crafty Worker Is a Happy Worker

When I was at a thrift store recently, I came across a collection of over 100 booklets from the 1950s that had been, as stated on their back covers, "Prepared especially for the GM Men and Women by the General Motors Information Rack Service."

The booklets covered an astoundingly wide variety of topics, including craft-related titles such as *Transformagic: How to Make Old Furniture into New, You Can Make Art Your Hobby, Easy Patterns: How to Sew with Only Simple Pieces of Material,* and my personal favorite, *There's Magic in Clay.*

Can you imagine any major corporation today handing out booklets to its workers titled *There's Magic in Clay*? In this slim volume, potter Kay Harrison teaches how to make a half dozen clay figures, including a clown, which "makes a very good conversation-piece when the neighbors call, and builds up your ego to no end," and "Jug Head," an anthropomorphic liquor bottle whom Harrison assures her readers will be a "cunning woman's home companion on the nights that papa is 'out with the boys.'" (One can't help but wonder if Harrison's magic was in the jug rather than the clay.)

The instructional utility of the booklets vary widely from volume to volume. The sewing book, for instance, provides detailed instructions for making garments such as a romper sunsuit, a pinafore, and a cuffed box jacket.

The art book, on the other hand, glosses over four years of art school in a dozen pages with nearly useless advice such as "Don't worry about perspective … Learn it by drawing!"

Even the silliest of the books are a joy to read. I'm sorry large companies are no longer encouraging their employees to enjoy life, but I'm glad that in the last several years, people have rediscovered the joy of crafting on their own. Long live the DIY Information Rack Service! ✂

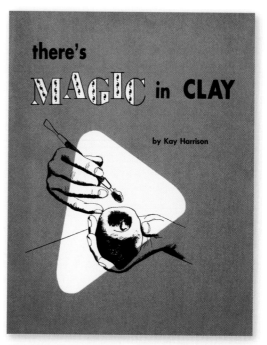

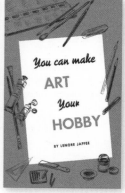

The craft booklets shown here are just a few of the hundreds that were made available at no charge to GM employees in the days of yore.

The Whitey-Tighty Wallet*

An underwear wallet might not be the first thing that jumps into your mind when you're planning your next craft project, but you might consider it after you see this clever how-to and its fabulous result.

You will need: Size 4 or bigger little boy's underwear, one package bias binding, ¼ yard of 45" coordinating fabric, medium- or craft-weight fusible interfacing, seam ripper, scissors, iron, optional doodads (velcro, buttons, snaps, clear vinyl for an ID window, perhaps)

1. Cut

a. Using a seam ripper, remove the elastic waistband and leg bands.
b. Cut eight 4.5"×8" rectangles. Cut 1 using the front of the underwear, 1 from the back, and 4 from a matching cotton fabric. Also cut 2 from mid-weight fusible interfacing.
c. Iron interfacing on both pieces of undies.

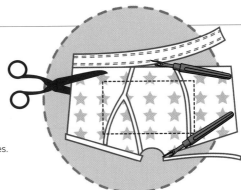

2. Sew

a. Customize 2 of the cotton pieces with pockets and spaces for credit cards using more cotton, velcro, buttons, and snaps. You can even add an ID pocket using 20-gauge clear vinyl.
b. Sew the pieces together along their long sides to make the inside and outside of the wallet. For the inside, sew 1 of the "customized" cotton pieces to 2 plain cotton pieces. For the outside, sew the front of the undies to the back of the undies and sew them both to the remaining "customized" piece. Iron flat.

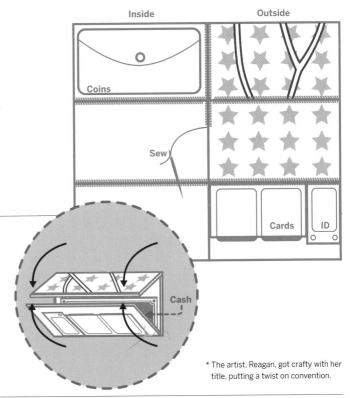

Inside   Outside

Coins

Sew

Cards   ID

3. Finish

a. With right sides out, zigzag front and back together along all edges.
b. Bias bind all around the edges.
c. Fold the bottom flap up once and whipstitch to create the bill pocket. Ta-da!

Cash

* The artist, Reagan, got crafty with her title, putting a twist on convention.

Reagan Copeland is a coffee-fueled high school senior living in Pennsylvania. Her favorite color is grey, and her adoration for sewing, knitting, and doodling is rivaled only by her love of the theatre.

Illustrations by Dave McMahon

OUR FAVORITE
TRINKETS & TREASURES

1. Glass Bottom View

» This vibrant stained glass panel incorporates the bottoms of wine bottles, as well as a painted glass scene of Dionysus and Maenad.
annabuilt.com/sg_oneofakind.html#

2. Sand Castles Made Simple

» That's the title of the book that includes this photo. Since when do simple sand castles look like the set of Harry Potter 5?
sandyfeet.com/sand/cheese/

3. Three Feet Under

» Made from vintage yardsticks, this planter is just one of many repurposed crafted items found at
happyhomedesigns.com.

4. Knitted "Hallowig"

» Perfect for those bad hair days.

craftzine.com/go/ hallowig

5. Paper Pottery

» Artist Brian Jewett uses carnival tickets rather than clay to make his bowls.

craftzine.com/go/ jewett

6. Rustic Crochet Hooks

» Carved from tree branches and then waxed, these hooks take handmade to a whole new level.

craftzine.com/go/ rustichooks

7. In Record Time

» Of all the cool crafts made from vinyl records, this 45 rpm clock, with laser-cut numbers, aluminum hands, and a plastic pop-out center, tops our list.

vinylux.net

8. Sew Darn Cute!

» Anyone into Blythe dolls will go gaga over the one-of-a-kind handmade dolly duds offered at this site.

sewdarncute.com

9. Coziest Pipes

» Rebecca Vaughan knitted this cozy cover for her ceiling pipes, as well as her linoleum table, heater, and even her entire minivan. So much for teapots!

knitknit.net/cosies.html

10. Floppy Disk Blank Book

» Just a decade ago, a floppy disk wasn't anything more than an ugly information holder. Now, used as a cover for this 50-page notebook, it's suddenly retro, nostalgic, and full of charm.
scrapsofpaper.net

11. Who Done It?

» Jim Rosenau makes custom shelves around any theme, but be patient — it can take years before he finds just the right books.
thisintothat.com

12. Wearable Breakfast

» Fortify your day with this fried egg patch. Accompanying fork and knife pins included.
tokkisom.com/shop_access.html

13. Go Pac-Man!

» This beaded strand of power pills, impish ghosts, and a frantic Pac-Man were made by glass beadmaker Sarah Hornik, who's a web designer by day, crafter at night.
glassbysarah.com

CRAFTER

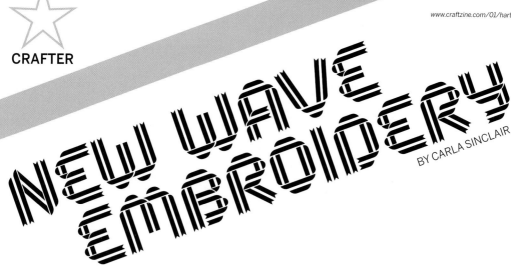

NEW WAVE EMBROIDERY

BY CARLA SINCLAIR

Sublime stitcher Jenny Hart thinks outside the hoop.

Six years ago, embroidery was still stuck in the granny section of the craft world, under the same roof that housed lace doilies, crocheted tea cozies, and needlepoint roses. Then along came spunky Jenny Hart, with her stitched celebrity portraits, pin-up girls, and pink-flossed skulls, and a new wave of embroiderers was born.

Hart, a 30-something who lives in Austin, grew up reading underground comics, enjoying the illustrations as much as the stories. "It was this interest in illustration that was the driving force behind me getting into embroidery."

Embroidery depends on illustrations, she explains, and this is what makes it different from something like knitting. "That's why it appeals to me so much."

Before Hart became a rock star in the world of floss and hoops, she was an exhibition coordinator assistant, helping to hang lights and install shows. "Then I had the incredible good fortune of being laid off. I was elated. I thought, 'This means my dream can come true.'"

Hart wasted no time diving into her craft. After her first embroidery project — a portrait of her mother — "I got totally addicted to it. I had to be embroidering something every day for the next 3 or 4 years." From family portraits, she made the jump to pop stars, who include Marianne Faithfull, Dolly Parton, and Iggy Pop.

Many portraits later, Hart decided to try decorative embroidery, but was disappointed in the tired patterns and outdated instructions that were available. It was all too rigid and uptight.

Instead of dropping the craft, she decided to create her own patterns, such as a girl skull, tiki head, sleek cat, hula dancer, cherries, and an electric guitar. She also created her own line of hip embroidery kits called Sublime Stitching, which contain original patterns, floss, a hoop, and instructions, including a few of her own invented stitches. "I like to invent stitches and give them funny names, and they're often derivative of other stitches that are really basic." Some of her original stitch names include the Bamboo Stitch, the Twinkle Stitch, and the Scalloping Chain.

She's even got her own way of tying things up. "There's a technique I use for making a knot at the back when you're finished that my students [Hart teaches embroidery workshops in New York] were calling my 'Secret Knot,' which I thought was funny."

Although Sublime Stitching is "overwhelmingly successful" (thanks in part to Plaid/Bucilla, a major craft manufacturer who recently picked up her kits, which means Sublime Stitching can now be found in Michaels, Jo-Ann Fabrics, and the like), she's only just scratched the surface of what she wants to do. "I've always wanted to see embroidery not just in nontraditional subject matter, but also with nontraditional materials."

For instance, she embroidered a screen door, using bright green and yellow glow-in-the-dark 6-strand floss, which she doubled over to give it

"My interpretation of the mythic, Mexican weeping woman — as a chola with a tattooed tear and her cry of 'Donda está mi hijo' tattooed on her forearm."

thickness. The design included the word *Entre* surrounded by curlicues. "It wasn't a project you cuddle up with in an armchair," she says, referring to the awkward size and weight of her "canvas." "It was more like a hardware project."

Another off-the-wall project Hart recently completed was her leather lace-embroidered skateboard. For this, she drilled holes where the needle would normally intersect the fabric, and used feather, whip, running, and straight stitches to decorate the underside of the deck. Last spring, the board was auctioned off as part of a benefit to fund a new skate park in Knoxville, Tenn.

Perhaps one of Jenny's more bizarre projects is *Oh Unicorn*, a portrait of a man kissing a unicorn-horned woman, stitched on deerskin canvas, using human hair for thread.

Hart used her own hair as well as human hair extensions she bought at a wig shop. Soft deerskin, which she compares to a shammy, was a lot more difficult to embroider than she'd predicted. "I was really naïve. I thought this was going to be a wonderful, beautiful experience, but it wasn't easy."

To decorate her skateboard, Jenny Hart uses leather lace for the border. In traditional embroidery, this stitch is known as a running stitch.

So what does Hart see on her artistic horizon? She says she wants to embroider a giant sculpture to give people another perspective. "It's not about making a pretty object, but more about getting people to look at it really, really close up to see the essential elements of it."

Without giving away too much, a couple of her big ideas include embroidering a freestanding wall with enormous rope, and embroidering a huge panel of metal with wires. "I really want to do a big honkin' piece. I want to spend a good year or two working on something large." She says there's something hidden about the finer details of embroidery, and with these large-scaled pieces, she hopes to unveil — and at the same time magnify — some of its secrets. ✕

▣ *To embroider your own skateboard, see page 121.*

Carla Sinclair is the editor-in-chief of CRAFT magazine.

SCULPTING TRASH

BY MICHAEL SHAPIRO

From discarded water dispensers to rusty wheelbarrows, sculptor Patrick Amiot can turn anything into fabulous art.

Patrick Amiot's urban folk art, as he calls it, has transformed his entire neighborhood. In 2001, Amiot was a struggling ceramic artist in Sebastopol, Calif. Depressed about his flagging career, he decided to try something different.

Gathering some old vacuum cleaners lying around his house, a used barbecue, and a wheelbarrow, he fashioned these items into a gigantic fisherman, finishing it off with loud vibrant paint. He then displayed his quirky creation in his front yard. "I thought the city would come down on me, but people came out of the woodwork and told me how much they loved it."

Feeling inspired, he next created his version of the Statue of Liberty, made with junk, including a blender, car parts, a garbage can, and his wife's yellow nylons. He put it on display on Sept. 10, 2001. "The next day, after the terrorist attacks, I put a sign on it reading, 'In memory of all the innocent people who died on Sept. 11, 2001.'"

Next, Amiot built a fireman using a vintage San Francisco garbage can for the body, antique car horns for ears, and water dispensers for both the head and helmet. He offered it to the fireman living across the street, who happily accepted. Then almost everyone on the street wanted a piece.

Amiot now has 25 brightly painted, larger-than-life art pieces that grace at least half of the houses on Florence Street, where he lives. The street is as surreal as an outdoor funhouse.

Almost everything in Amiot's wildly creative sculptures is recycled, salvaged, or donated, and much of it has a strong connection to the local community. When Sebastopol's Barlow Company,

an apple cannery, went out of business, Amiot was given 45,000 #8 size lids, each about the size of a small dinner plate. Hundreds of these lids make up the owl's body in the *Owl and the Pussycat*, a whimsical piece that also includes recycled steel and old car parts. The canoe that the owl and pussycat ride in was a decaying boat rescued from the nearby Russian River.

When asked how he makes his art, Amiot is virtually speechless. It's not that he's secretive — it's just not something he can easily explain.

Pressed about his technique, Amiot waxes philosophical: "Making sculpture from raw materials can be expensive, complicated, and time-consuming. Just slapping it together is simpler and more efficient. You have to content yourself with what's around. If I lived in the middle of the desert, I'd make sand sculptures."

The secret to putting all these pieces together is self-tapping screws and a welder. "It's simple to weld," he says. Lastly, he affixes everything to the base with the self-tapping screws.

Part of Amiot's appeal is that he doesn't try to make his art "perfect." The beauty is in its imperfection, its uniqueness, its earthy irregularity. "Some guys can't help but be anal, but I let the viewers use their imagination," he says. "I look back to when I was grinding metal or chiseling stone — all this fighting to control an element. Now I let the elements control me." ✕

MAKE contributor Michael Shapiro was an editor for O'Reilly's Global Network Navigator in 1994-95, and is the author of the award-winning *A Sense of Place: Great Travel Writers Talk About Their Craft, Lives, and Inspiration*.

Photography by Christopher Lucas

★
**NEW LIFE TO
CAST ASIDE GOODS:**
Part of Amiot's appeal
is that he doesn't try to
make his art "perfect."
The beauty is in its
imperfection, uniqueness,
and earthy irregularity.

WORLD ON A STRING

BY XENI JARDIN

Liza Lou builds intricate scenes one bead at a time.

Beads adorn ordinary tools in traditional cultures: rattles, pouches, baskets of dried grain. But through her beadwork, Los Angeles-based artist Liza Lou constructs life-sized environments that feel anything but ordinary.

Like virtual worlds online rendered one pixel at a time, Lou's parallel realms grow bead by bead, often taking years to complete.

Her career as the world's most extreme fine-art beader didn't take shape overnight, either. Born in 1969 in New York, Lou moved to California to study painting at the San Francisco Art Institute, but a trip to a Bay Area bead shop one day changed her plans. Paint and canvas soon gave way to three-dimensional surfaces that absorbed beads like stippled brushstrokes.

For *Kitchen* (1995), Lou collected household appliances, then covered them in multicolored bead strands. Papier-mâché forms became cereal boxes and fruit pies, and beads covered every surface: a sky-blue sink, a refrigerator, even lowly dust balls on the floor. The 168-square-foot installation took Lou five years to finish, but its completion instantly transformed her into an art star.

Hundreds of volunteer beaders helped Lou stitch *Backyard* (1997), a 525-square-foot suburban yard with a peyote-stitch garden hose, a beaded Weber grill, and 250,000 blades of faux grass. For *Star-Spangled Presidents* (2001), Lou beaded 42 black-and-white presidential portraits, each wrapped in a frame of gold beads. In 2002, she was awarded a MacArthur fellowship.

Her most recent exhibit, which includes *Cell*, a beaded death row prison cell, and *Security Fence*, complete with razor wire and human figures in states of agony, was installed at London's White Cube gallery last spring. "It's a response to what's going on in the world around terror, and a reflection on the idea of containment," she explains.

"Beadwork is a slow, quiet practice that counts the hours, not unlike someone doing time. Sometimes when I'm working on projects, I wake up and realize, 'Wow, eight hours have gone by, and I've only finished two inches.' That narrowness of moment-to-moment focus can drive you insane."

In addition to her studio in L.A.'s artsy Topanga Canyon, Lou also maintains a workshop in Durban, South Africa, where she commissions assistance on large-scale installations from local Zulu beadworkers.

"I was planning my next piece, and realized — if I do this on my own, it will take 75 years. I needed manpower, and I was fascinated with the rich history of Zulu beadwork, and it all came together."

Lou rented a dance hall in Durban in 2005, and began connecting with local craftswomen. They soon found they had something more in common than the mere mechanics of beading.

"Labor is undervalued everywhere, and the laborer too often ends up discarded," said Lou. "So much of this craft is about taking pride in doing something well.

If you make something concrete as well as you possibly can, you can't be tossed away as easily by the world. It becomes its own merit, its own

A close-up of the 525-square-foot *Backyard* reveals excruciating detail.

★ Liza Lou's social commentary *Trailer* (1999-2000), detail shown here, paints a vivid and haunting tale. Lou faithfully covered the interior of a vintage aluminum Airstream trailer strictly in subdued but sparkling grays and browns.

dignity. It becomes a source of pride."

For Lou, watching how people responded to the London show underscored something significant about the work itself.

"I'd see people walk into the gallery, and some might glance around for a few minutes, then walk out. Nothing out of the ordinary about that, until you stop and realize that they were looking at a piece that took three, four, maybe five years to complete."

Lou says the protracted, bead-by-bead development of her massive works is not unlike the pace of human life.

"We hope to leave something of ourselves behind, something of value from our labor. But even that's not guaranteed, no matter how hard you work or how beautiful the results might be. There might be a war, an earthquake, a flood. There are no promises that anything of you will be left."

Despite the obvious difference between low-tech beading and high-tech coding, people who labor over digital tasks may find something in common with Lou's craft.

At the heart of beadwork, from peyote stitch to Zulu strands, there is math. Patterns of pixels on screen aren't all that different from patterns of beads on threads, when closely observed. For both kinds of crafters, careful planning and methodical, cyclical thought are required. In each, the whole is dependent on the precise sum of its parts, and the tiniest calculation errors can botch the most mammoth of projects.

Coders and beaders might share something more existential, too, says Lou.

"I've become more and more pessimistic over time, and I don't expect what I create to last forever. I think it's better to value the daily aspect of what you do — to find joy in daily work — instead of laboring for some great, final reward."

Lou's latest exhibit took place in Japan earlier this year, at the Museum of Contemporary Art in Tokyo. ✕

Xeni Jardin is a tech culture journalist and co-editor of boingboing.net. She was once an enthusiastic peyote-stitcher and beader before she discovered pixels.

GEEK GIRL TAKES ON FASHION

BY NATALIE ZEE

Designer Diana Eng creates style with a scientific twist.

Fashion designer Diana Eng, self-proclaimed fashion nerd, says "I'm trying to make fashion and technology something people can relate to in their lives." She graced the nation's TV waves earlier this year as a contestant on Bravo's *Project Runway*. "Part of the challenge is that I'm just curious. I *feel* like it can definitely be done, but I want to *know* if it can be done."

Diana Eng is our kind of fashion designer. She loves science magazines and comic books. She's a techno-geek at heart who codes her own HTML on her blog (dianaeng.com) because she likes having design control over the layout and images. She loves to experiment and figure out how things work. It is with that curious spirit that Eng, who graduated from the Rhode Island School of Design (RISD) in apparel design two summers ago, may be our new hope for technology's future in fashion.

It's no surprise that Eng's equal love of fashion and technology goes all the way back to her childhood roots in Jacksonville, Fla. There, with her computer scientist father and architect mother, Eng and her younger brother (now an MIT student) would frequent art shows and museums, especially works by artist Richard Meyer. "We always had design books around," she remembers.

In the science world, her mentor was her grandmother, a math and computer science teacher, who would take Eng as a young girl to math conferences and who introduced her to research on spirolaterals.

By the time she reached her teens, Eng was already a guest lecturer at the Florida Council of Teachers of Mathematics annual conference, speaking on the benefits of using visual aids, such as origami or spirolaterals, while teaching math.

Her fashion designs show the beauty in the technicalities of math and science. Her signature showcase piece, the *Inflatable Dress*, was created in school with her classmate and now business partner/co-designer, Emily Albinski. The concept for the project was to explore how design changes through shape and color.

The dress' center is a modified hand vacuum connected with tubing to help inflate the dress. Varying the amount of air from the vacuum changes the silhouette of the dress. The idea is simple: "You might be at prom and you want to be the only person with your dress. So, if your dress changes shape, you will always be the only person with your dress," Eng concludes with a girlish laugh.

Her next piece, the *Heartbeat Hoodie*, takes technology a bit further. It's a stylishly designed hooded vest jacket complete with camera and heart monitor. The concept explores the idea of involuntarily documenting parts of one's life at moments of interest or

Diana Eng is our kind of designer. She loves science mags, comic books, and codes all her own HTML.

Illustration by Jemma Hostetler

excitement. Strategically placed above the eyes on the hoodie, the camera takes photos as the wearer's heart rate increases. The camera itself is wired discreetly through the seaming of the garment to a BASIC Stamp that communicates with a wireless heart rate monitor.

The BASIC Stamp uses an algorithm to analyze the heartbeat for increases that might signify a moment of excitement, as opposed to physical exercise. Since the photos are taken involuntarily, the wearer may discover new points of interest they may not have been conscious of.

In contrast, Eng's *Mathematical Knits* are a beautiful collection of sweaters and scarves that debuted at MIT's Seamless fashion show earlier this year. The technology behind the aesthetic is much more subtle. Created on Eng's large knitting machine, each design is based on the mathematical Fibonacci sequence, a series of numbers relating to

★ *Biomimetic Clothing* (left) converts from one look to another based on principles from biomimetics, deployable structures, and TRIZ, the Russian theory of inventive problem solving. The *Mathematical Knit Scarf* (right) uses number patterns to dictate how its decorative structures are knit.

the laws of nature and the golden mean.

Just as Leonardo da Vinci and the other Renaissance artists were exploring the harmony of proportional mathematical aesthetics through their paintings and sculpture, Eng came across her ideas in much the same way. "I noticed that there is a certain technique [on the knitting machine] that I like to use, called the holding technique because you have to be constantly calculating the stitches in your head," Eng explains. "The way it works is you add one number and add it to the second number and then you add it to the third number, which is very similar to the Fibonacci sequence. I figured

Photography by James Duncan Davidson

I should just use the Fibonacci sequence and see how that turns out."

What's next for Eng? She and Albinski have formed Black Box Nation (blackboxnation.com), a new company forging ahead with their collective vision of fashion and technology. The line debuted at the Maker Faire fashion show in April, along with their stylish collection of jewelry made from electronic parts.

Eng describes the meaning behind the name: "We are using the term 'black box,' which is the technical term for electronics where you don't know what's inside," she says. "Nation is added because we feel that, a lot of times in the United States, people don't know what's going on; they just know that such and such will happen, but they don't know what makes it work."

The discourse has already begun. Models backstage at her Maker Faire show were overheard

★ **PCB Radio Jewelry (left), by Emily Albinski, is comprised of a handmade circuit board with FM radio, earrings as speakers, and a connecting wire bracelet. When the vacuum is activated, the *Inflatable Dress* (right) grows to a bell-shape with tendril-like spikes on the back and large pillows of air on the sides.**

talking about their outfits, how they were made, and the technologies behind them. Most asked questions and seemed interested in learning more. Show attendees oohed and aahed in unison as the stories behind the pieces unfolded.

Diana Eng's dream of moving technology into the fashion mainstream is starting to happen, as we see her become one of the defining leaders in this new realm of fashion technology. ✂

Natalie Zee is associate editor of CRAFT and writes for the craft blog at craftzine.com.

70s Tie-Upholstered Switch Plate

A polyester classic, the ultra-wide, double-thick necktie of the 1970s never came back in style like it should've. Nevertheless, you can still pay homage to one of the Me Decade's most garish moments in men's fashion with this upholstered switch plate.

You will need: Wide 70s necktie, plastic light-switch plate, scissors, pencil, contact cement with brush, X-Acto blade, needlenose pliers or tweezers

1. Cutting

Using scissors, cut the bottom 8" from the necktie. Cut any seam along the back. Remove the lining and superfluous filler.

Place the fabric face down, and using the widest part of the tie, center the plate face up and trace its placement. Cutting ¾" outside your traced lines, create a 6"×4¼" rectangle.

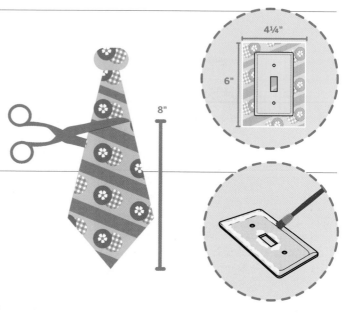

2. Gluing

In a well-ventilated area, coat the back-side of the fabric and the front side of the switch plate with contact cement (include the inside edges of the switch opening). Allow both pieces to dry.

On the backside of the switch plate, brush a ½" trim on the inside edges, and ¼" around the outside edge of the switch opening with contact cement. Allow to dry.

Place the coated necktie fabric face down on a flat surface. Place the switch plate face down, centered, onto the fabric, pressing to make full contact.

3. Finishing

Using an X-Acto knife, cut a diagonal "V" in each corner of the fabric, and an "X" in the center of the switch opening. Do not cut clear to the edge of the switch plate — allow for a ⅛" buffer.

Pull each edge of the excess trim over each side of the switch plate, and press into place. Using needlenose pliers or tweezers, pull the small triangular tabs through and over the edges of the switch opening, pressing each one into place.

Using the tip of the X-Acto blade, puncture a slit in the fabric centered directly over each screw opening. Mount plate to wall as usual.

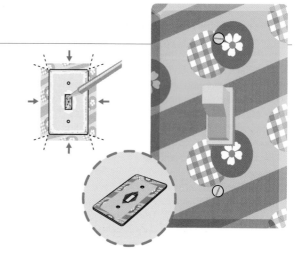

Matt Maranian is a bestselling writer, designer, and bon vivant whose books include *PAD* and *PAD Parties*. He lives in New England.

Illustrations by Dave McMahon

Ulla-Maaria Mutanen
Linkages

>> Ulla-Maaria is the developer of Thinglink (thinglink.org), author of the HobbyPrincess blog (hobbyprincess.com), and researcher of design and innovation work at the University of Helsinki. ulla@aula.cc

Crafter Economics

In his essay "The Long Tail," *Wired's* Chris Anderson wrote, "The future of entertainment is in the millions of niche markets at the shallow end of the bitstream." I've been wondering how the economics change when exchange moves farther down the tail from manufactured products to crafts.

When I post a picture of a self-made laptop sleeve on my weblog, does that mean that my laptop sleeve is now "on the market"? Many of us would probably say that the answer depends on whether I offer to sell my sleeve for a price.

Well, let's assume I do not, but then a stranger emails me offering to swap it for something she has made. And when my friends see my creation, it inspires them to make their own versions that inspire yet more crafters. Some sell theirs on Etsy (etsy.com).

In this story, the self-made laptop sleeve on my blog is clearly part of some system of exchange. Still, it seems that in this system of makers, users, buyers, and sellers, the idea of the market is broader than classical economics would have it. We are used to thinking that:

» profit motivates exchange,
» exchange is based on money,
» price is determined by supply and demand,
» and demand can be purchased (stimulated through marketing).

But my goal wasn't to profit. Money did not always change hands, there was no need to negotiate a price except on Etsy, and nothing was spent on marketing.

If we still want to call this system a market, we need to update the definition of the term.

First, learning, recognition, and reciprocity motivate crafter exchange at least as much as economic profit. The American Association of Hobby Industries reports that only 15% of crafters claim an interest in selling their creations. The rest have other reasons for making things.

Second, barter and conversation are more important modes of exchange to crafters than monetary transactions. I impart bits of my know-how in exchange for your recognition and advice. I may also swap my laptop sleeve for something that you have made. To sell my laptop sleeve for money is just one option among many.

Third, links determine the crafter value of an object. Crafters value objects that can teach them something new. A product rich with stories about its origin, maker, materials, and techniques of manufacture is infinitely more interesting than a product without a history. For a crafter, a product without links is a dead product.

Lastly, crafter demand shows as recognition, not just purchases. Recognition depends more on recommendations than marketing dollars spent on media space. Companies who have dabbled in purchasing recommendations will keep burning their fingers.

So is there a different crafter economics? I think there is, and the crafter exchange logic is moving up the tail to manufactured goods. Witness the growing preference for customized fashion, cars, and electronics. In America, the size of the craft and hobby industry has risen by 50%, from USD $20 billion in 2000 to USD $29 billion in 2004. Although money doesn't move crafters, crafters are moving increasing amounts of money. ✂

Zombies
&
Robots
&
Bears

OH MY!

BY MARY BELTON

The world of creepy
cute dolls and the
people who make them.

K lops is a sexy beast, with curved pointy horns and a bottom row of sharp jagged teeth. His handsome felt-lined button eye catches mine, and I admire his polyester strip of man-fur that runs from his chin to below his navel. He's the ultimate Cyclops, a lady's monster, if you will, at least according to his creators at Creature Co-op (creatureco-op.com), who run a booth at the annual Los Angeles Bazaar Bizarre, a bustling craft fair that sells everything from knitted food to laminated purses to magnet boards made with vintage frames. I get to the fair right after the doors open, and already it's mobbed with people. As I make my way down the long aisles of vendors, I realize that hand-stitched dolls — which range from adorable Japanese-influenced animals to half-crazed fuzzy creatures that hover somewhere between cute and creepy — are attracting swarms of people and seem to be one of the hottest categories at the fair.

According to Robert Kalin of Etsy (etsy.com), an online marketplace for handmade stuff, the stitched doll trend started in 2003 with the popularity of Uglydolls (uglydolls.com). Six years earlier, Sun-Min Kim, just out of art school, stitched her first Uglydoll (based on a drawing by her boyfriend, designer David Horvath) named Wage, a pointy-toothed, orange plush creature who wears a blue apron. According to his back story, Wage works at Super Mart, even though management at the store doesn't know of Wage's existence, and when Wage is off-duty, he tries to make friends with fire hydrants, phone booths, and anything else that looks lonely.

Kim and Horvath's first order was with Giant Robot, an Asian-Americana pop-culture shop in West L.A., who asked for 20 Uglydolls. Those sold out in one day, so Giant Robot ordered 20 more, then 50. The order shot up to 1,500 dolls within 18 months, and Kim's hands began bleeding from sew-

> *The order shot up to 1,500 dolls within 18 months, and Kim's hands began bleeding from sewing them all herself.*

ing them all herself. That's when she decided it was time to hire help. By 2003, Uglydolls were a bona fide hit with college students and soccer moms alike. These days, 60,000 Uglydolls are manufactured each month in China, and hand-stitching dolls is a crafting craze.

Hundreds of crafters use the internet as their trading post, where they buy, sell, trade, and exchange information on these weird little dolls. Communities of crafters are sharing their creations and feeding off of each other's ideas, generating mini-trends like knitted bears in striped sweaters, knitted Yodas and zombies, cats with two heads, hand-stitched body parts, or crocheted cupcakes.

Etsy recently hosted a contest for a specific style of stitched doll known as *amigurumi*. Amigurumi is a Japanese word meaning crocheted or knitted doll. The jury is out among doll makers as to whether the cuteness and weirdness of these creatures are part of the definition, but Kalin, organizer of Etsy's amigurumi contest, encourages entrants to design "mashups and mutants." Contest entries included butter & toast, a pirate squid, a turtle bird, an

angry sheep, and a coffee bean. The pirate squid took first place.

Kalin says amigurumi dolls are the most popular items on the Etsy site, with a sell rate of over 50%. They go for anywhere from $10 to $100, but with certificates of authenticity, Kalin believes the toys could be big business, claiming, "Amigurumi could be the next Beanie Baby, only handmade."

The most popular seller on Etsy is a crafter known as Gourmet Amigurumi, a.k.a. Beth Doherty, whose own site includes bikini-clad bears, squishy bean-filled snails, beatnik cats, and screaming punk girls. A painter with a fine arts degree, Doherty took up crocheting to unwind after she was hospitalized with crushing migraine headaches. When she found links to Japanese amigurumi sites on craftster.org, she began making small crocheted animals. "First I made a dumpy little cat," she says, but eventually graduated to little girls and snails. Doherty admits, "I'm embarrassed to tell people what I do. I have to explain that they aren't hideous."

In fact, it's just the opposite. Because of her attention to detail and exquisite craftsmanship, many other doll makers cite Doherty as an inspiration. A perfectionist, Doherty admires Japanese artisans for their precision. "Japanese crafters don't mind spending lots of time on a small piece, and it's those little details that really appeal to me."

Doherty also claims Andy Warhol among her influences. "His work isn't my favorite, but I admire his attitude toward art. He blended consumer goods with art." Doherty seems to have a knack for this as well. Kalin says that not only are Doherty's dolls the best sellers on Etsy, they are usually sold within 20 minutes of being uploaded.

Guam-born Jess Hutchison, who now lives in San Francisco and is known for her hundreds of knitted robots and cute creatures, doesn't like to sell her goods. "I prefer that people ask me how to make a toy rather than how to buy it. Politically, it's a natural extension of the DIY thing." Hutchison explains, "We want to make stuff as humans, and it makes

Knitted zombies and their victims (preceding pages) are made by Hannah Simpson, who's obsessed with George Romero's 1978 Dawn of the Dead. Influenced by Japanese toys, these robots (right) were knitted by Jess Hutchison, who's been making toys since she was a kid. Cute with an edge, crocheted rabbit and lamb dolls (page 44) by Beth Doherty are examples of what she sells on her site gourmetamigurumi.com.

sense making stuffed toys, creating something functional in an emotional way, something that has an artistic quality to it."

Unlike Doherty, Hutchison didn't study art in school. During a period of unemployment, her sister taught her to knit. Once she got tired of making sweaters, she thought, "What if I knit a robot?"

Hutchison counts Disney concept artist Mary Blair among her influences. Blair was the concept artist for several Disney films, including *Alice in Wonderland*, *Cinderella*, and *Peter Pan*, and was the designer of the Disneyland attraction It's a Small World. Her colorful, quirky abstract work inspires many stitched doll makers.

Hutchison makes frequent pilgrimages to

> "*We want to make stuff as humans, and it makes sense making stuffed toys, creating something functional in an emotional way.*"

Japanese bookstores for amigurumi pattern books. "They are doing the most innovative stuff," she says, "and the patterns are visual, much simpler than the way we write them. You don't need to know Japanese. There are new pattern books every week."

Hannah Simpson, a British pub worker in Oxford, England, gets 10,000 hits a week on the Flickr page displaying her knitted *Dawn of the Dead* dolls. A novice, Simpson started knitting dolls a little over a year ago. Obsessed with George Romero's 1978 *Dawn of the Dead*, to the point where Simpson owns four copies of the horror flick and has seen it at least 50 times, the zombies were her first stab at doll making. Why the love affair with *Dawn of the Dead*? Simpson calmly explains, "They're us, we're them," referring to the film's underlying message of consumerism, conformity, and primal fear.

Once Simpson began knitting dolls, she couldn't stop. In a constant rush to make more, she knits on her day shift at the pub and brings her wool everywhere she goes. Her latest project is knitting the

band The White Stripes. Next, she plans to recreate characters from the *Star Wars* movies of the 70s.

German artist Patricia Waller straddles the fence between kitsch and art with her crocheted creations that are better fit to sit in a gallery than to be tossed on a sofa. Citing Marcel DuChamp, René Magritte, Jeffrey Koons, and Vincent van Gogh among her influences, Waller's pieces include co-joined teddy bears, a shark clenching a human leg between its teeth, and dentures soaking in a glass jar.

To Waller, crocheting is a political statement. She contends that wool is consistently neglected in the art world and says, "Of course I take advantage of the image of 'housewife art,' so that, at first glance, my works appear innocent. On a closer look, however, people will discover a sort of vicious irony. I am interested in transferring the material to another level."

Australian toy artist Jäke Henzler has also become obsessed with making knitted dolls. Before picking up the needles, Henzler's primary creative outlet was digital imaging, but no longer. Henzler confessed, "While I'm knitting one toy, I get an idea for the next one and start it right after I'm finished."

Henzler's pieces are both funny and thought-provoking. A homage to the Van Eyke masterpiece *Arnolfini and His Bride* features knitted versions of Arnolfini, his bride, and the baby that Henzler imagines the couple was expecting in the original portrait. The knitted baby is still attached to the bride by a red wool umbilical cord. A set of eight knitted tampons replete with strings and tiny faces is Henzler's comment on feminine product advertising. Vicki Vom Vom is a red doll vomiting brown yarn with green flecks. Celine, yet another creation, holds the heart that has been ripped out of her chest, in her hands.

Lately, Henzler has changed his approach, complaining, "I'm tired of the creepy thing for now." Instead, he has decided to go for straight-up cute. Among his latest creations are Trevor the Effeminate Koala and a one-eyed polar bear named Sasha.

Cute. But kinda creepy. ✄

➕ *To crochet your own amigurumi robot, see page 123.*

Mary Belton is a freelance television producer. She lives in Los Angeles and spends her free time seeking out the odd and fantastic.

Crafting GEOMETRY

BY ARWEN O'REILLY

What coral reefs, chaos, and the bias cut have in common.

Dr. Daina Taimina strides up to the podium in an elaborate crocheted skirt. She's speaking at the Gathering for Gardner, an invitation-only mathematics event in honor of Martin Gardner, the famed mathematician and puzzler (he wrote a column in *Scientific American* for almost 50 years).

Her talk is about crocheting math, a subject that has garnered her increasing media attention in recent years. Much of the audience is well-versed in the geometry of the hyperbolic plane, but most people haven't had a chance to play with some of the best hyperbolic models to come out of academia: Taimina has crocheted them.

Models existed before, but they were fairly fragile. She explains: "William Thurston made a paper and tape model around 1986. In 1997, I was going to teach a class and thought, 'I can't use this; it's falling apart!' So I decided to crochet one."

With the mathematical programming inherent in a crochet pattern, she was able to make a sturdy, pliable model that really gets across the physicality of the hyperbolic plane, even to ordinary observers. Her students went nuts.

"It's not that you haven't seen it before," she points out. "Here are pictures from my garden." Slides flash by of ruffled lettuce, curly kale, the rippled edges of a sea slug. "Well, the sea slug is not from my garden," she laughs. She goes on to explain what geometrists already know: while a sphere has constant positive curvature, a hyperbolic plane is the opposite of a sphere, having constant negative curvature. It is always curving away from itself, causing the ripple we see in nature and in math textbooks.

Having the physical models on hand allowed Taimina to explore new ideas and prove new con-nections. "The other realization is that maybe this can be a fashion line for mathematicians!" she says, twirling her skirt. She gets a big laugh and some eager looks. Mathematical fashion may not have hit the big time yet, but the intersection of crafting and science is definitely on the rise.

Margaret Wertheim, for one, is thrilled about it. A science journalist, she founded the Institute For Figuring in 2003 with the idea of drawing attention to what she calls the "poetic and aesthetic aspects" of science and mathematics. Taimina and her hus-band, geometrician David Henderson, were among the first people invited to lecture. Ever since then, the Institute has been tackling topics like knot theory, tensegrity structures, and paper folding, as well as continuing to explore hyperbolic crochet. They coor-dinate lectures and exhibitions, drawing the attention of the art and design worlds to objects like Taimina's.

The Institute has a number of exhibitions coming up in the next year (as well as supporting her with a grant), including the display of a collaborative crocheted coral reef at the LACE gallery. Responding to Taimina's work with hyperbolic space, Wertheim and her sister Christine, a co-director of the Institute, started playing around and crocheting their own forms, including shapes that looked like kelp and cactuses. "We had them arranged on our dining room table," says Wertheim, an Australian, "and as they grew, we realized that it looked like a coral reef." A crocheted reef is no coincidence; like kelp, coral reefs have hyperbolic geometry in them.

The sisters started inviting people to contribute pieces of the reef, and soon had so many submissions that they ended up in the position of curating cro-chet. In the process of looking for a space to exhibit, they were surprised and delighted to find that a number of other textile artists were also making coral reefs independently. The show is tentatively titled

This amazing crocheted Lorenz manifold took Dr. Hinke Osinga 85 hours and 25,511 stitches to complete. ≫

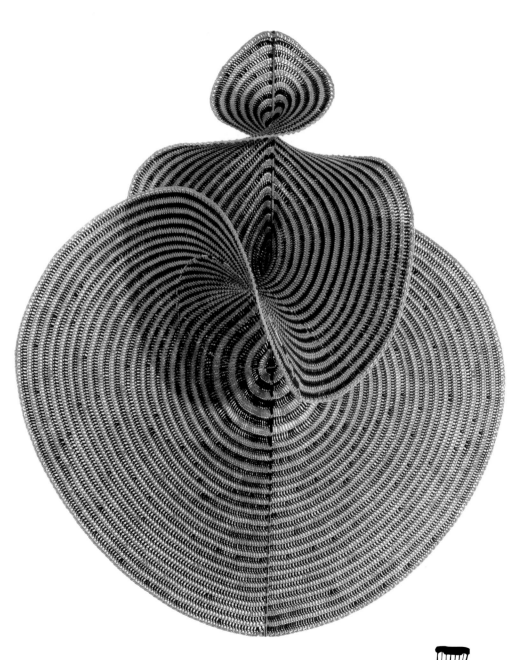

47
FEATURE

"I've Got a Coral Reef, Too" and will include different artists' interpretations.

All of a sudden, craft, math, and art are coming together in ways that no one could have predicted. Or, as Wertheim puts it, "crafts have become mechanisms for making really interesting art." On the other side of the Atlantic, University of Bristol researchers Drs. Hinke Osinga and Bernd Krauskopf are finding out the same thing. Like Taimina, Osinga originally crocheted in her spare time, but she and Krauskopf realized that they could apply it to their research. The result was a crocheted Lorenz manifold, which is a representation of chaos in the famous Lorenz system.

The Lorenz manifold is an example of a complicated surface that shows how chaos arises in a system that changes in time. Osinga and Krauskopf

> *"It was important to us that anyone should be able to repeat our creation."*

develop computer methods to find and visualize such surfaces. (Osinga gives an example: "Such a surface may consist of all possible positions and velocities of a spaceship such that it reaches a specific point." The work has applications in laser dynamics and neurological research as well.) While their work centers on computer methods, the crochet project was "driven by the need to see and feel the real thing ... we realize now how much artistic value the crocheted Lorenz manifold has."

The Lorenz project turned out to be an inspiration for artists and a great public outreach tool. It's striking how eagerly these mathematicians share their processes, true to the open source roots of crafting. "It was important to us that anyone should be able to repeat our creation, but we were actually worried that nobody would try," admits Osinga, so they offered a bottle of champagne to the first person to crochet another manifold.

She shouldn't have been nervous; they got three responses in two weeks. "We learnt how the use of handcraft for visualization makes complex mathematics extremely accessible to the general public," she says. "It serves as an eye-opener to people who thought mathematics wasn't for them."

Drawing more people into geometry is something Dr. Sarah-Marie Belcastro welcomes. A young, energetic assistant professor at Smith College and the co-director of the Hampshire College Summer Studies in Mathematics program, she bewails the state of university geometry programs. "We're lucky if a given institution has one undergraduate geometry course," she points out. Her homepage, toroidalsnark.net, has a fascinating collection of links to examples of mathematical knitting, crocheting, and fiber arts, and is one of the best resources on the state of this emerging field.

Belcastro knits things most people only vaguely remember from high school or college math classes: three-holed toruses, Klein bottles, and projective planes. Her Klein bottle hat looks like something surrealist fashion designer Elsa Schiaparelli would have sent out on the runway, and she points out that math influences fashion more than we may think. Every time you slip on a bias-cut dress, you're paying homage to Madeleine Vionnet, who invented cutting across the grain of fabric for superior drape and considered herself a geometrician.

The confluence of fashion and math is making waves in the book world, too. Belcastro and her crafting partner are working on a proposal for a book on mathematics and fiber arts, aimed at crafters and mathematicians. *Knitting Nature*, a book by Norah Gaughan, was just published this past summer by Stewart, Talbori and Chang, and features skirts, sweaters, shawls, and shrugs inspired by starfish, tortoise shells, and honeycombs. (The intersection of knitting and math seems to be "out there in the air," as she puts it.) Next year will see the arrival of a long-awaited book of patterns edited by Sabrina Gschwandtner, the founder of knitknit.net, an art and crafting zine that has printed patterns for geodesic hats and ASCII weaving.

While all these books may have seemed completely on the fringe a few years ago, the timing may be just right. Geometry has arrived. Craft is cool again. And maybe those math classes will start filling up now. ✄

+ For more images and resources, see craftzine.com/go/geometry.

My Brain on Acid, *a huge orange symmetrical hyperbolic plane, 30cm in diameter, by Christine Wertheim of the Institute For Figuring.* »

Arwen O'Reilly is staff editor of CRAFT.

Photograph by Anitra Menning/IFF Archive

Long Live
GOCCO

BY JILL BLISS

Can we resurrect the world's coolest printing machine?

I love the idea of screen printing, but never quite mastered the processes of emulsion, darkrooms, exposure times, power washers, proper ink consistency, and proper tilt of the squeegee. So, when a few years ago a Japanese artist friend suggested I try her screen printing system brought from Japan, I was a little leery. "No, it's easy," she insisted.

She described Print Gocco as a self-contained unit that both exposes and prints on paper cards, combining the basic principles of screen printing and rubber stamping. That was all I needed to pique my interest and troll online for a kit of my own.

Although there were a few high-priced listings on eBay, after an extensive online search, I was able to secure a kit from an art store's dusty and forgotten shelves for approximately $150. Once my Gocco arrived, it took less than an hour to make my first set of prints: the "whirl wheels" flower design on the back of old postcards.

Unlike screen printing, there are only four basic steps to the Print Gocco system: make your design, expose the artwork inside the machine using bright flashbulbs, ink the screen, and print using the screen inside the machine.

Like all Gocco owners, I showed off my first set of prints and kit to everyone I knew. And, like so many other Gocco geeks, I was even able to convince my local art store to stock the system and its accessories, as well as offer classes. During my initial Gocco online foray, I also found the 609-member-strong Yahoo Gocco group, a treasure trove of Gocco information, ideas, and swaps. There is also the more recent Flickr Gocco group, with 189 members.

Gocco has been a staple in Japan since the 1970s, when it was first marketed as a print kit that the whole family could enjoy. In the past few years,

interest in the Gocco system has been steadily increasing in the United States, as artists, designers, zinesters, and craftsters discover it, fueled by online word-of-mouth. Enthusiasm for Gocco is contagious. The universal first response to seeing a Gocco in action is "I gotta get one of these!"

It's no surprise to Gocco enthusiasts that nearly one-third of Japanese households own a Print Gocco, using them regularly to print personalized holiday cards or invitations. What is surprising is the lack of great success for Print Gocco outside Japan, save for our own legion of devoted followers.

With all this growing interest, you'd think Gocco's parent company, Riso, would increase production, marketing, and sales in the United States, but on the contrary, Riso decided to pull the plug on us shortly before last Christmas. The machines have been discontinued, with limited supplies and accessories on hand. U.S. suppliers have already run out of stock on the familiar powder-blue B-6 kit, and prices for them are skyrocketing on eBay.

So I wonder, can Gocco be saved? No one is certain. With the discontinuation of Gocco here, many worldwide fans speculate about the future of the system in their own countries, and Riso remains tight-lipped about any future plans. My own fansite, savegocco.com, attempts to keep atop the latest Gocco news and the efforts to save it. Once my site receives 1,000 sign-ins in the guestbook, I'll approach other companies who may be interested in manufacturing it. Fingers crossed and go go Gocco! ✂

Jill Bliss' comfort with both handcrafts and technology can be traced to her tech-tinkerer dad, who was a member of the Homebrew Computer Club, and her mom, who made and sold crafts each Christmas. Visit her site at blissen.com.

Clockwise from top left:
Jill Bliss, Evan B. Harris,
Apak, Onsmith Comics, Kate
Bingamon, David Lu, Cole
Johnson, Ian Lynam, and
Mizna Wado.

Many of these works are
limited edition prints
highlighted in "We Heart
Gocco," a show organized
by the Wurst Gallery in
Portland, Ore., intended both
to extol the delights of Gocco
printing and draw awareness
to its potential demise.

For more GOCCO images,
see *thewurstgallery.com.*

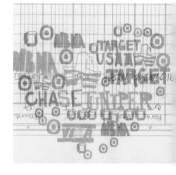

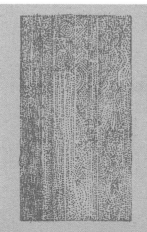

Susie Bright

Susie's Home Ec

» Susie Bright is an amateur dressmaker and a professional writer. She blogs at susiebright.com.

The Case of the Missing Curve

There's a secret in the fabric shop. Dozens of women come in all day to purchase soft minky for baby blankets, retro oilcloth for totebags, and Betty Boop flannel for pillowcases.

But these aren't just eager homemakers and doting moms. These are seamstresses who are terrified to make something to wear for themselves. They're hiding out in Home Dec, because the last time they made a dress, it was such a disaster they're still trembling with shame.

What was the problem? Why did they never attempt a jacket, a pair of jeans — a T-shirt, even? The answer is in one dreaded word: curves — the diabolical design of the female figure. Aside from all the adolescent angst they cause, curving bodies present the first real challenge to the home stitcher.

The chief culprits are found in the unstated prerogatives of the most well-known pattern companies. They don't advertise this fact, but Vogue, Butterick, Simplicity, McCall's, and New Look all design their patterns for a B-cup, no matter how large the chest measurement. You could cut out the tissue with a 40" bust, and it would still be a B-cup; this is astounding when you think of how few women actually fit that measurement.

These same patterns also place the apex of the bustline (where your nipples sit) at a high angle, pointing at the stars. If you're more at sea level or below, the finished garment is going to look ridiculous.

Why do they do this? To be honest, designing for the flattest plane is the simplest, most forgiving assignment. It also allows for the most variety. When they say that slender, boyish figures look the

best in high fashion, it's not a lie, or even a prejudice. A tremendous variety of clothes look great on hangers. If you have that hanger-like look, you should try every crazy outfit under the sun, because it will probably look swell on you.

Curves require more care, and more discrimination as to what will flatter. They take longer to draw, to cut, to sew — they are simply more labor-intensive.

Have you noticed how bridal designs in recent years have focused on sleeveless, strapless gowns? They're all the rage, and it's not just because they're sexy or romantic. No, it's because drawing and tailoring a *sleeve* (a huge curve if there ever was one) would add a tremendous cost to the dress, and so designers like Vera Wang realized that the profit margin lay in cutting those darn things off.

Is there any hope? Yes, and a well-drawn pattern by someone who has devoted themselves to curves is the place to begin. Don't reinvent the wheel unless you're enrolling at Parsons and need the exercise.

Popular patterns from companies like Burda, Hot Patterns, and Kwik Sew all feature designs that assume that as one gets bigger in size, the cup size should change with the chest measurement. Simplicity just offered a couple of experimental patterns where they offer different cup sizes (one trendy jacket and one shirt-dress), and I intend to flood them with encouraging letters.

And then there's *Fit for Real People*, a manual as unpretentious as it sounds. It shows you how to slash, spread, and tape any pattern tissue into submission! You'll be flabbergasted by how great the models look whether they're 6 feet tall or 80 years old, top heavy, pear shaped, or round as a barrel — all because they finally have something that fits. ✕

PROJECTS

Photograph by Jason Madara

For crafts you can wear, convert an unwanted pair of shoes into chic knitted slouch boots, then top them off with a tank top that flashes animated programs. For home hacking, create a plexiglass room divider that looks like a humongous ant farm, and construct a swanky pad for your feline royalty. Then step outside and build a feng-shui-friendly, Jet Age bamboo and rock garden.

THE ELECTRIC TANK TOP

By Leah Buechley

USE SILVER-COATED THREAD AND A MICROPROCESSOR TO MAKE PROGRAMMABLE LED CLOTHING

» The Westin Shanghai has a majestic LED staircase complete with its own sound and light show created by Color Kinetics.

▶▶ I built this shirt to experiment with wearable computing and electronic technology, and realized along the way that the basic materials were actually quite easy to work with.

There's lots of room for creativity at all levels, so I was inspired to write this do-it-yourself guide. Although challenging, everyone should play with this stuff! It's great fun for both geeks and divas — build a sparkly fashion accessory and program it with hacker animations. Even better, make it a group project.

The code I wrote starts a "glider" (a figure in the Game of Life universe) that marches around your garment forever. Play with the code to get other life patterns. You're guaranteed to turn heads whenever you're out on the town.

» For the last 30 million years, butterflies have used the same method as LEDs for emitting light.

» For *Star Wars: Episode II, Attack of the Clones* (2002), the LED lights on the back of the clone troopers helmets displayed "THX 1138," paying homage to director George Lucas' famous student movie.

» In the early 1970s, red LEDs were used in the first digital watches like this Hamilton Pulsar.

For links to these and other related stories, check out craftzine.com/01/led.

Leah Buechley is a Ph.D. student in computer science and a member of the Craft Technology Group at the University of Colorado at Boulder, where she has found a place that she can unite all her interests. "I get to play with computers and sewing machines, electronics, fabrics, and beads: heaven!"

Illustrations by Tim Lillis

WHAT YOU'LL NEED

[A] Silver-coated thread
members.shaw.ca/ubik/
thread/order.html

[B] Surface mount LEDs
100, or as many as you'd like. I
used a super-intensity red LED
(Digi-Key part #67-1695-1-ND).

[C] AVR ATmega16 microcontroller Digi-Key part
#ATMEGA16L-8PC-ND

[D] IC socket for your
microcontroller Digi-Key
part #A9440-ND

[E] AVR programmer
Digi-Key part #ATSTK500-ND

[F] 9-15VDC power supply
For your STK500 program-
mer. Available at RadioShack.

[G] USB serial adapter and
included software This will
attach your programmer to
your computer.

[H] Battery and holder
I used a standard 6V camera
battery. For the holder, use
Digi-Key part #108KK-ND.

[I] On/off switch Digi-Key
part #401-1000-1-ND

[J] 30-watt or higher soldering iron and lead-free
solder Remember: Keep it
lead-free!

[K] Multimeter

[L] T square or ruler

[M] Assortment of silver
and brass crimping beads
At least twice as many as you
have LEDs.

[N] Garment or a piece of
fabric and a pattern

[O] Sewing needle, fabric
marker, bottle of fabric glue

[P] Scissors

[Q] Sewing machine

[R] Insulating backing
fabric (not shown)

**For additional info
about materials
and equipment, visit
craftzine.com/01/led.**

Photography by Leah Buechley

➤➤ CREATE A
BLINKING TANK TOP

Time: **One Week** Complexity: **Extreme**

1. DESIGN YOUR PATTERN

1a. Decide on the number of LEDs you want and their general placement. I decided to sew a simple tank top, and I chose to place the LEDs evenly across my top every 2". Since my tank top is approximately 28" around and 12" tall, I needed 84 LEDs.

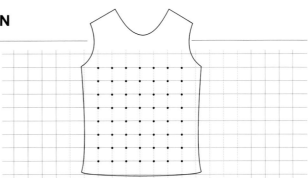

2. MAKE SEQUINS WITH YOUR LEDS AND BEADS

If you're not up for soldering, you can substitute traditional through-hole LEDs for the surface mount LEDs, twisting their leads into spirals to make them stitchable. Read about this technique at craftzine.com/01/led.

2a. Get crimping beads and surface mount your LEDs. Using a soldering iron with a very clean tip, place the tip of the iron into a bead. Melt some solder onto the outside of the bead. With the soldering iron, drag the bead up to the LED as shown (middle right). When the melted solder touches the LED's contact, the bead will adhere to the LED. Lift the soldering iron out of the bead.

Now take some measures to distinguish the cathode lead (-) from the anode lead (+) of each LED. The cathode end is often marked with a green line on the front or back of the surface mount package. To distinguish the two, solder a brass crimping bead to the cathode lead and a silver bead to the anode lead for each LED.

2b. Solder beads to the leads for your battery and switch, so that they can also be sewn on. This is the switch sequin.

TIP: If your soldering iron tip is dirty, it will stick to the bead and make the job very difficult. If this is happening, you should clean or replace your tip.

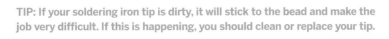

3. SEW YOUR LED PATTERN

3a. With a marking pen, mark the lines for your LED pattern on the garment. Also, mark where you want your microcontroller (IC socket) and power supply to be. You want a grid of conductive traces where the vertical traces do not touch the horizontal ones. A simple way to do this is to put one trace on one side of the fabric and the other trace on the flip side of the fabric, utilizing the fabric as a natural insulator. The lines for the vertical traces should be on one side of your garment, and the lines for the horizontal traces should be on the other.

I marked both sets of lines on both sides of my tank top to make sure my lines were well placed. Use a T square to get good right angles and straight lines.

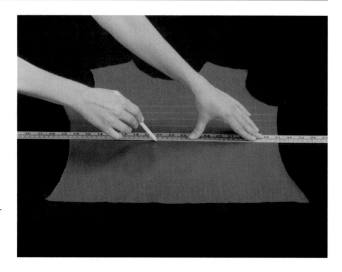

3b. Make a bobbin of silver-coated thread for your sewing machine, and put it in the machine. Use a spool of ordinary thread for the top thread.

Using silver-coated thread in the bobbin of a sewing machine will allow you to sew conductive horizontal traces on one side of your garment and conductive vertical traces on the other side. As you sew, the bobbin thread will remain on the underside of the fabric you are sewing.

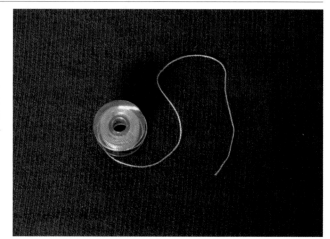

Q: So, if I want conducting vertical traces on the back, where should I draw my lines?

A: You should draw them on the opposite side, the front. The lines you draw will be facing up as you sew so that you can follow them. The conductive trace, from the bobbin thread, will be on the opposite side of the fabric.

3c. Sew 1 trial row-column crossing, and use the multimeter to make sure your threads are being sufficiently insulated by the fabric. If your fabric is too thin, the bobbin thread may be pulled through the fabric, and your crossing traces may short out (*see related sidebar on page 62*).

If there is contact at your intersections, you will need to take action to correct this. As you are sewing out the traces, you should stop the sewing machine just before each intersection, and, without breaking the threads, move your fabric past the intersection and resume sewing. This will insure that the silver-coated thread stays on the proper side of the fabric at each crossing.

3d. Sew out your vertical traces. Flip your garment over and sew out your horizontal traces.

You should stop your pattern stitches a few inches from the IC socket to leave room for the knots you will make when sewing the socket on by hand.

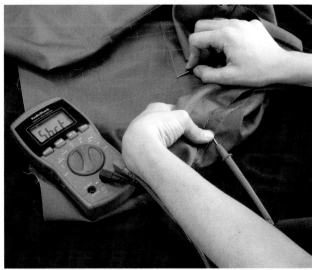

NOTE: Here are the top and bottom views of my partially assembled tank top after I sewed on my traces.

TOP

BOTTOM

4. PREPARE AND SEW ON THE IC SOCKET

4a. Before you start sewing threads onto the IC socket, you should familiarize yourself with the pins of the microcontroller. Follow along on the pin layout diagram for the ATmega16 chip, shown below.

All the pins labeled PA0-PA7, PB0-PB7, PC0-PC7, and PD0-PD7 are general-purpose input/output pins that can be used to power LEDs and the like. See my sample code and header files at craftzine.com/01/led to see how to reference and control individual pins with your code. You can download AVR microcontroller datasheets from craftzine.com/go/avr8.

4b. Use the following diagram as a guide to tell you which thread goes to which socket. (If you've chosen a different number of rows or columns, you'll assign them to the PA-PD pings somewhat differently.)

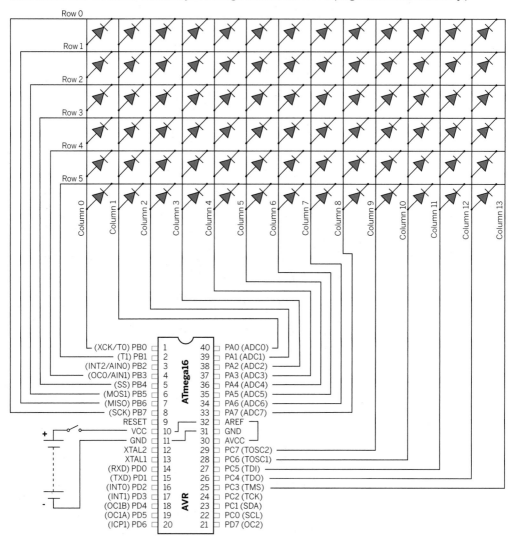

▶▶ "I added an infrared receiver to it and my friend wrote a program for his PalmPilot that lets people beam patterns to it"

4c. Trim the pins off the bottom of the socket and pull off any tape or other material blocking the holes. If necessary, drill out the holes so that a needle can pass through them. Position the socket where you want it on your garment and stitch it in place with silver-coated thread, sewing traces from each microcontroller socket to the pattern traces you sewed.

You want to make sure that the silver-coated thread makes contact with each socket hole, but also be careful that no two threads cross. This is a delicate job that requires some patience, but if you're used to doing soldering or any other meticulous work, it should be no problem.

4d. Make sure that you tie your knots where there is ample room for them (away from the socket) and where they're less likely to cause shorts with neighboring traces. Coat each knot with fabric glue. This will keep knots from fraying and coming untied.

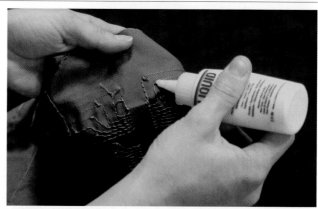

 Q: What is a short?

A: A short or "short circuit" occurs when the positive terminal of a power supply is connected directly to the negative terminal of a power supply. On your shirt, if 2 neighboring traces are touching while one of them is high (positive) and the other is low (negative), a short circuit is created. This kind of short circuit will prevent your LEDs from lighting up and is likely to cause your microcontroller to overheat and eventually die. Short circuits in more high-powered applications can cause fires and explosions.

5. SEW ON YOUR LEDS

5a. Attach the cathode end of each LED to a row, and the anode end of each LED to a column (or vice versa). If you did not take steps during the soldering phase to differentiate the cathode from anode leads, you will have to make the distinction now.

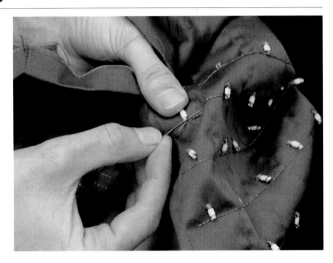

The cathode end of the LED is often marked with a green line on the front or back of the surface mount package. If you are able to find this marking despite your soldering, you can use it. Otherwise, learn to distinguish the direction from the appearance of the face of the LED. Test one by running a current through it for reference. Be careful to use a voltage and current appropriate for your LED.

5b. While sewing, take care to make good connections between your thread and each bead, looping the thread through each bead several times, as shown here.

The fastest way to sew is to stitch each row and column continuously, not stopping to tie off the thread for each LED. In other words, sew in the cathode end of one LED, and sew down your row to the next LED cathode without cutting your thread. However, this makes replacing badly sewn or broken LEDs harder, since you'll have to cut the continuous thread and tie the ends off in the event of a problem.

Alternatively, you can sew each LED on individually. This will make repairs easier, but your sewing will take much longer. I chose the first option for faster sewing, but I did have to replace a few LEDs.

Q: Can I do anything to make sure my LEDs won't break off?

A: **Before you sew any of the LEDs, abuse them a little to test your solder joints: twist and tug on the beads. The weaker joints will break and you'll be left with the sturdy LEDs. I tried this method on the second shirt I made and not one of my LEDs has broken since I sewed it, and they've even withstood a few washings.**

6. TEST YOUR CIRCUIT

6a. Using a multimeter, make sure none of your traces are shorting out with one another, and all of them are leading to the appropriate LED rows and columns. Silver-coated thread tends to fray and give off small "hairs."

Make sure there are no miniscule conducting hairs interfering with any of your traces.

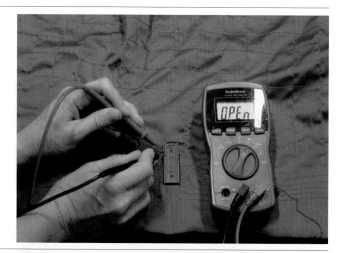

6b. You may also want to make sure your LED pattern is working properly by attaching the leads of your power supply to the rows and columns of your pattern in turn.

Look at the specifications that came with your LEDs if you're not sure what power supply to use or you may fry all of your LEDs!

My multimeter (in beep mode) doubles as a low-current power supply, illuminating an LED when its leads are attached to the right threads.

6c. Once you've done some thorough testing, glue an insulating backing onto the traces you sewed for your IC socket, so that your power supply will be easy to attach and these traces will remain in place without fraying with wear.

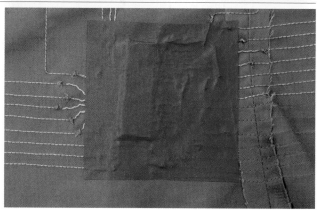

7. ATTACH POWER SUPPLY AND SWITCH

7a. Sew the switch and power supply to the garment.

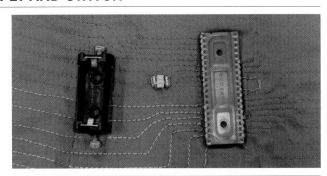

7b. Glue an insulating backing over your power supply and switch traces so that you will not accidentally turn on your display.

Here is the inside view of my tank top after I sewed on the power supply. Notice the insulating backing that was applied prior to sewing.

You're done! Now program your micro-controller (see craftzine.com/01/led for coding examples) and you'll be the light of the party.

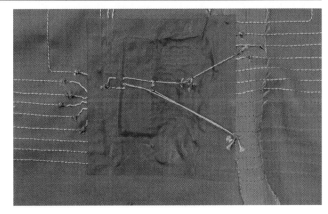

FINISH X

MORE RESOURCES

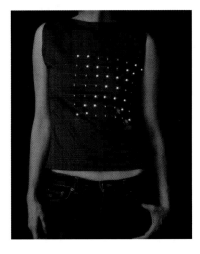

This shirt is a big attention-getter. People want to touch it and scrunch it up and examine it up close. Everyone wants to know how to wash it (carefully and by hand!) and whether it will work in the rain (alas, no). I've gotten all sorts of business advice, from the cynical, "They'll be making those in China for $5 a piece," to the exuberant, "You're going to make millions!"

It's been especially fun to wear since I added an infrared receiver to it and my friend wrote a program for his PalmPilot that lets people beam patterns to it. People have a blast with this, though I imagine it can be potentially risky to relinquish control over what you're wearing!

For tips on troubleshooting and customizing your garment, options for a simpler project, and how to make an electronic sewing kit, visit craftzine.com/01/led.

Now go out and wear it!

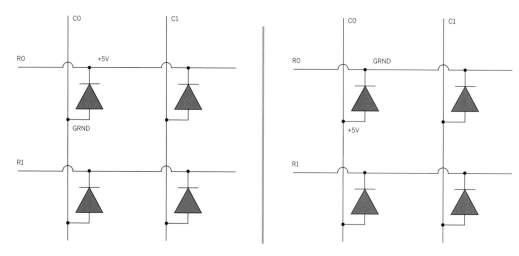

Two diagrams of a row-column LED pattern.   **Fig. A**   **Fig. B**

DESIGN AN LED PATTERN

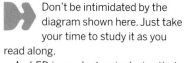 Don't be intimidated by the diagram shown here. Just take your time to study it as you read along.

An LED is an electronic device that emits light when you apply a current to it. Unlike an ordinary light bulb, however, diodes allow current to flow in only one direction. That's a good thing for us, because we can use this property to power a lot of LEDs with just a few electrical connections coming from the pins of the microcontroller we'll be using.

The pattern shown above will supply power to X number of LEDs, using 2 times the square root of X number pins. For example, you can power 100 LEDs with 20 pins. As you can see from the diagram, the LEDs are arranged in rows and columns with the anode (+) end of each LED attached to a row and the cathode (-) end of each LED attached to a column.

Each row and column is then attached to a pin, allowing the microcontroller to control each LED individually. Suppose we want only the LED at row 0 column 0 (LED R0 C0) turned on.

To accomplish this, we first turn all of the LEDs off by setting all of the rows to ground and all of the columns to +5 volts, applying a reverse voltage to all of the LEDs. Then, to turn on LED R0 C0, we set R0 to +5V and C0 to ground. LED R0 C0 is the only LED with current running through it, so it will emit light. This type of design allows us to control each LED individually, but does not give us complete flexibility.

For example, it is impossible to simultaneously turn on only LED R0 C0 and LED R1 C1, because it would also turn on the other two LEDs in the pattern.

But that's not a problem for us. We can still flash complex patterns and animations by exploiting a biological phenomenon known as persistence of vision.

To make it appear as though LED R0 C0 and LED R1 C1 are on at the same time, we quickly flash LED R0 C0 first and then LED R1 C1, and then repeat this cycle for as long as we want the illusion to appear. As long as our eyes can't detect the flicker (and they can't in our project), we perceive only the diagonal line of light.

Fig. A: With C0 set to ground and R0 set to +5V, current can't flow through LED R0 C0.

Fig. B: With C0 set to +5V and R0 set to ground, current flows through LED R0 C0.

NOTE: The row/column wires do not actually contact one another at the intersections.

LEDs are everywhere! Here are more inspirational examples of LEDs incorporated in wearable fashion and designs for life.

» Lumiloop is a reactive bracelet that interprets the motions of the wrist and generates illuminated patterns in response. mintymonkey.com/lumiloop_p1.html

» The Glo LED necklace emits a brilliant shine inside a 10mm Swarovski crystal cube. craftzine.com/go/ledglo

» Issey Miyake illuminated vest. enlighted.com/pages/imvest.shtml

» SparkLab's Wearable Light bracelet. sparklab.la/bracelet.htm

» Janet Hansen created this faux fur vest with LEDs that pulse in various patterns. craftzine.com/go/furled

» LED Throwies! craftzine.com/go/throwies

» Spoke POV: PacMan LED bike wheel images. craftzine.com/go/spokepov

JET AGE GARDEN

By Mister Jalopy

ASIAN-INSPIRED SANCTUARY

With $100 and a spare weekend, you can build a corner garden that will impress your granny and offend local Japanese garden enthusiasts. A profoundly meaningful art form, the Japanese garden should be studied over a lifetime. Like playing the game Go, you can learn the rules, but very few people will ever understand it in a significant way. Perhaps you've already dedicated your life to some other equally rewarding pursuit. Well, then I can offer the low-rent 1960s suburban cul-de-sac Danish modern via *Sunset* magazine by way of Home Depot garden with a vague Asian influence. It would be criminal to call it a Japanese garden, so let's call it the Jet Age Garden to be sure no one gets hurt.

» The classic tiki mug grew in popularity in the 50s and 60s, and is now the perfect kitschy accessory for any backyard garden party.
en.wikipedia.org/wiki/Tiki_mugs

» The tsukubai is found in Japanese tea gardens to "purify" oneself by resting a moment and washing your hands.
craftzine.com/go/tsukubai

» The yukimi, a popular stone lantern in Japanese gardens, is also called the "snow-viewing lantern" because its roof captures the falling snow so well.
cherryblossomgardens.com/alanterns2.htm

» Apple CEO Steve Jobs adheres to the Japanese Zen aesthetic as seen in the visual elegance of Apple's products.
craftzine.com/go/zenjobs

Mister Jalopy breaks the unbroken, repairs the irreparable, and explores the mechanical world at hooptyrides.com

Illustrations by Tim Lillis

WHAT YOU'LL NEED

MATERIALS AND TOOLS

[A] 15" plastic play ball
like you find in the metal cage
at Target

[B] Bowling ball

[C] Plastic bucket

[D] Dry concrete mix

[E] Bottle of liquid
cement color

[F] Bamboo pole

[G] Copper tubing and
tubing cutter, ¼"

[H] Hose bib to ¼"
compression fitting

[I] Needle valve, ¼"
Make sure it comes with
2 compression nuts.

[J] Flashlight

[K] Liquid Nails glue

[L] High-gloss shellac

[M] Gravel (optional)

[N] Spade

[O] Hole saw

[P] Wrenches (2)

Photograph by Meiko Arquillos

▶▶ CREATE YOUR OWN GARDEN SANCTUARY

Time: **A Weekend** Complexity: **Easy**

1. MAKE THE CONCRETE BASIN

The basin collects the water from the dripping bamboo fountain. Rather than buy some sort of receptacle from the local gardening shop, cast your own basin in your own color and to your own specifications. Soon you'll start looking at cookie sheets and Tupperware as concrete molds to make table tops, stepping stones, and more.

1a. Create the mold by cutting a hole in a 15" play ball large enough to squeeze a junk bowling ball inside. I probably should have bought the SpongeBob ball as I felt awful cutting Pooh's face.

1b. With the bowling ball inside the mold, pour in water to get a sense of how much concrete will be needed. How much is enough? About 3" looked good to me.

1c. Pour the water from the mold into a bucket and mark the water height. That gives you a pretty good estimate of the concrete required.

Photography by Mister Jalopy

1d. Fill bucket with dry concrete mix to ½" below the line — about ¼ of a 50 lb. bag. That red detergent bottle is filled with 1½ quarts of water with the dye added. Add half the water, mix, add a little more water, mix, and keep adding water until all the concrete is wet like thick oatmeal. If water is sitting on top and won't mix in, you added too much water. Just sprinkle in a little more concrete.

NOTE: I added about ¼ of a bottle of charcoal dye to the water, but I should have added ½ a bottle, as I would have preferred the concrete to be darker.

1e. Remove the bowling ball from the plastic mold, then scoop concrete into the mold. You can dump any leftover concrete into a cardboard box lined with a plastic bag, just like the professionals. Or you can leave it in the bucket to solidify like the lazy people! In either case, throw hardened concrete away rather than trying to flush it down the toilet.

1f. Once the concrete is in the play ball mold, press the bowling ball in to create the center depression. Not going very well? Too wet? Take out the ball, add a little more concrete, mix, drop the bowling ball back in and allow to dry according to instructions on the concrete bag.

Follow the instructions to build the 1000% deluxe bamboo water dripper and cast concrete basin, then build the rest of your garden around it. A smooth river rock sea in a 4" deep "pool" lined with plastic, a couple of ferns, several clumps of mondo grass, and a couple of stepping stones complete the jet age look.

2. MOCK UP THE DRIPPER

The dripper is made from a length of bamboo that has been cut in two and connected to form an inverted L shape. A length of thin copper tubing, connected to a nearby water tap and outfitted with a valve to control the flow, is routed through the interior of the dripper.

2a. Before buying the bamboo, play with different arrangements of the major components to figure out how the garden should look and how much bamboo will be required. For my garden, the vertical upright, including subterranean portion, was 4' long, and the horizontal dripper arm was 22".

2b. Now it's time to buy the bamboo. By planning ahead, you can have your local bamboo shop cut to the correct size for you. If you arrive unprepared like me, you will need to hacksaw the bamboo at home.

2c. Refine your plans. Dig a hole about 1' deep and position the vertical upright to figure out final placement. Take measurements for the dripper arm length and mark the upright for drilling holes.

3. ASSEMBLE THE DRIPPER

3a. Using a pipe, knock the thin interior walls from the dripper arm. Do the same for the vertical upright, but don't knock the lid off!

NOTE: The top of the bamboo upright needs to remain closed to keep your dripper from filling up with rain and scary bugs.

3b. Measure the diameter of the dripper arm and use a hole saw of the same size to cut a hole in the vertical upright.

3c. Cut the tubing to length. Tubing cutters work like magic and are hard to beat for 10 bucks. Tighten, spin, tighten again, and more spinning will result in a perfect cut.

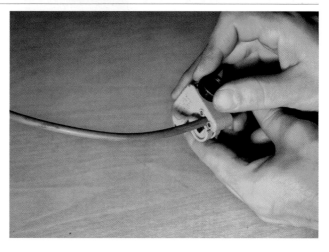

3d. Now we'll attach to the water source. Mercifully, there is a standard hose to ¼" compression adapter, so you can connect the dripper to any standard hose bib. Easy! Slide compression nut and sleeve over tubing. Then put the tubing end in the hose adapter. Tighten with 2 wrenches.

3e. Connect the needle valve. The ¼" needle valve comes with compression fittings that are tightened with the same 2-wrench technique.

3f. The needle valve will slow the water to a delightful, relaxing drip that would drive you insane if it were coming from the kitchen sink.

Drill a hole in back of the vertical upright and feed the tubing through, as seen in the illustration on page 73.

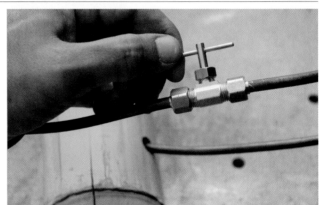

 Q: This seems like an expensive project. How much is it going to cost?

A: Actually, you can make it for under $100. The most expensive item is the bamboo at a budget-busting $45 while the bag of cement costs a mere $4. To save about $15, scavenge the automatic ice maker copper tubing and needle valve hookup from a trashed curbside refrigerator.

3g. Use a flashlight and thread the tubing through the interior of the vertical upright and then through the dripper arm.

3h. Glue, shellac, and concrete.

Glue the dripper arm seam with Liquid Nails, and then seal all the bamboo with high-gloss shellac.

With the dripper done, drop it in the hole and fill with gravel. Or if you have 2 sweet but unapologetic dogs like I do, mix up a batch of leftover concrete from the basin project.

FINISH ▣

MORE RESOURCES

Want to learn more about the profoundly moving and subtle art of real Japanese gardens? The following pragmatic and accessible Japanese gardening books will help you build a barrel bridge between your backyard and the temples of Kyoto.

Bamboo Source
Cane and Basket Supply
canebasket.com

Further Reading
Sunset Ideas for Japanese Gardens
by the editors of *Sunset* magazine and Sunset Books

Japanese Touch for Your Garden
by Kiyoshi Seike, Masanobu Kudo, Haruzo Ohashi

Japanese Garden Design (hardcover) by Marc P. Keane

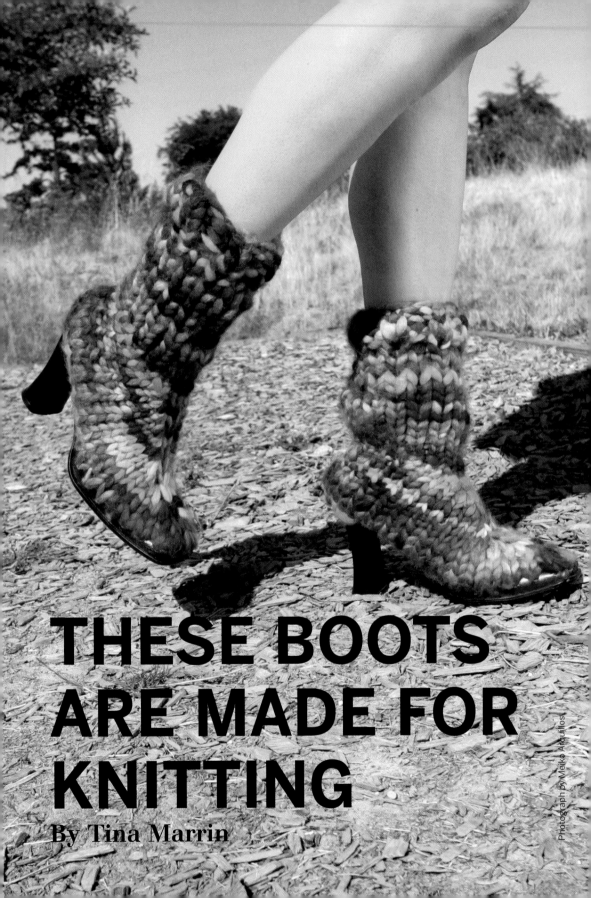

THESE BOOTS ARE MADE FOR KNITTING

By Tina Marrin

TRANSFORM BORING SHOES INTO AN AWESOME PAIR OF KNIT SLOUCH BOOTS

» The first time I saw a knitted boot in a magazine, I was inspired to try my hand at replicating it. The plan was to transform a pair of existing high-heel pumps (that I bought and rarely wore) into a pair of knitted, knee-high mukluks that I would always wear. I knew I needed to make evenly spaced holes around the base of the existing shoes to anchor the "cast-on" stitches, and that I would use a power drill to do it.

The bulky, rugged wool I chose provided an appealing contrast to the slender spiked heel. And as I rotated the shoe around, engulfing it in knitting, the process felt curiously sculptural – and more like what a potter would feel like at his spinning wheel as a wet lump of clay is being transformed into a fine vessel.

Tina Marrin, a native of Los Angeles, is addicted to knitting flirty braille sweaters, mermaid skirts, and tickle dresses (holes where the ticklish spots are). She also produced the permanent exhibition "Garden of Eden on Wheels: Collections from Mobile Home and Trailer Parks in Los Angeles County" at the Museum of Jurrassic Technology in Culver City.

WHAT YOU'LL NEED

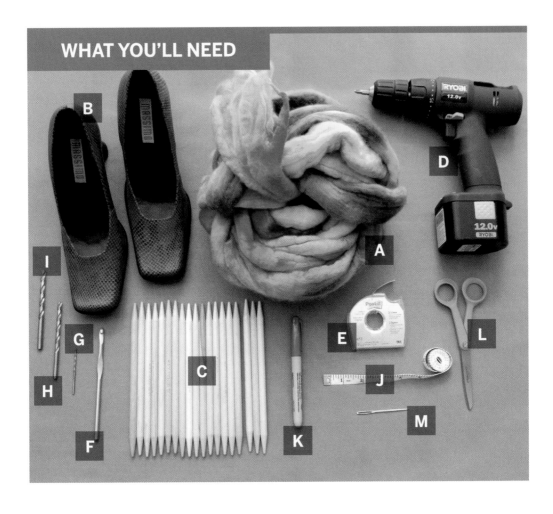

[A] 8 oz. merino roving (yarn) in "Autumn Forest" or color of choice

[B] 1 pair size 7½ Massimo "Aniliese" or shoe of choice

[C] Approximately 19 double-pointed needles U.S. size 13

[D] Power drill

[E] ⅜" wide Post-It brand cover-up tape (or other low-tack tape)

[F] Crochet hook U.S. size 10 or "J"

[G] ³⁄₃₂" drill bit

[H] ⁷⁄₃₂" drill bit

[I] ¼" drill bit

[J] Flexible measuring tape

[K] Permanent marking pen (or tack or push pin)

[L] Scissors

[M] Tapestry needle Size 13

Photography by Jack Marrin

▶▶ KNIT THOSE GROOVY BOOTS

Time: 2 Days Complexity: Medium

1. PREP THE YARN

1a. Take the 8 oz. ball of merino roving in the color of your choice and split the entire yardage into 2 strands by pulling apart, starting at one end, to make 2 balls.

NOTE: The roving naturally splits apart in 2 fairly equal parts, but you may need to coax evenness a little. Don't worry if it's not perfect — that's part of what gives the boot its texture and unique beauty. Roving is unspun wool and may seem fragile, but once knitted, it's quite strong.

1b. Split both balls of the roving apart again to make a total of 4 balls, and then again to make 8 balls. If the roving breaks apart, you can simply moisten it with saliva or water and vigorously rub the 2 overlapped ends together between your palms (creating heat) — this will permanently "felt" the roving together.

NOTE: At this point, I like to arrange the balls of roving according to thickness. I use the thinner roving for the beginning part of the project to avoid "fat feet."

Q: What if I want a fiber that looks more formal?

A: As long as you stick to a fiber that knits up at about the same gauge — 1 stitch (st) per inch — you can drill your holes the way the pattern describes. For thinner yarn, you will have to drill your holes a little closer together, and adjust the pattern for the new stitch count.

For more FAQs, go to **craftzine.com/01/knittedboots**.

2. MARK THE SHOES

2a. Run Post-It tape along the outside bottom part of each shoe upper, making a ⅜" tall "line" around the bottom of the shoe. Tear ½"–1" pieces of tape to conform to the curves of the shoe.

2b. On each shoe, mark the center of the toe box (eyeball it) just above the tape line.

2c. While holding the measuring tape up to the tape line, make a mark ¾" to the left of the center toe mark, just above the tape line.

2d. Continue marking every ¾" up to the center back seam. Don't mark on the back (or any) shoe seam. You now have 16 marks on one side of the shoe (not including the center mark).

2e. Repeat steps 2c and 2d on the right side. You should have 33 total marks (including the center toe mark) on each shoe. Remove the tape.

3. DRILL THE SHOES

3a. Drill a ³/₃₂" hole into each mark, beginning at the back of each shoe.

3b. Drill straight through the upper. These holes will be the "pilot holes." As you get to the toe area, be careful not to drill into the toe insole.

3c. Drill all marks on both shoes. Then replace the ³/₃₂" drill bit with a ⁷/₃₂" bit and re-drill all holes with the larger bit. Again, never drill into any seams.

3d. With a ¼" drill bit, re-drill all holes on both shoes. Dust off shoes.

NOTE: Take care not to drill into any insole or midsole materials. By creating the marks ⅜" above the bottom of the shoe upper, we have most likely avoided this possibility.

Also, be more careful with depressing the drill trigger when drilling with the larger bits, which "grab" more forcefully and quickly than the smaller bit.

4. CAST ON

4a. For each shoe, take a ball of roving and cast on (more like picking up stitches), beginning at the first hole to the left of the center back seam.

4b. Insert the crochet hook from the outside to the inside of the shoe. The yarn is waiting, on the inside of the shoe, to be grabbed by the crochet hook and pulled through (forming your first loop on the outside of the shoe).

4c. Slide the loop onto a size 13 DPN (double-pointed needle). Pick up (pull through) the next loop (left of the first loop) with the hook and place it on the same DPN. Don't pull too tight and put only 3-5 loops on each needle.

4d. Keep picking up, as before, until all holes have a loop coming out of them.

4e. Cut yarn. When cast-on is complete you'll have 2 tails (approximately 4½" each) inside of each shoe. We'll tie these tails in later.

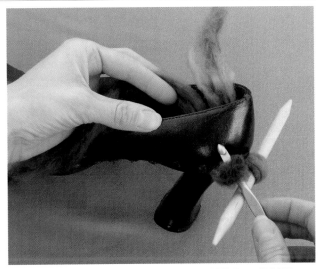

NOTE: I use 9 DPNs per shoe at cast-on stage.

5. KNIT THE SHOES

Row 1. Using a size 10 or "J" crochet hook and a ball of roving, "knit" the first row of stitches by pulling one stitch through with the crochet hook and place it (transfer) to a DPN. After the last stitch of the very first row is knit, cast on one stitch by making a firm backwards loop onto the right-hand needle. This is your center back stitch.

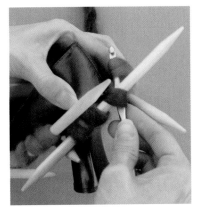

NOTE: The reason we are using a crochet hook for knitting the first few rows is because the stitches are very tight, since they are lodged between hard DPNs and a fairly non-pliable shoe. The stitches will eventually loosen up around row 2 or 3.

Row 2. k14, ssk, k1 (center toe stitch, or c.t.s.), k2tog, k15.

Row 3. k13, ssk, k1 (c.t.s.), k2tog, k14.

Row 4. k12, ssk, k1 (c.t.s.), k2tog, k13.

NOTE: *ssk* means slip, slip, knit and is a famous left-slanting decrease. *K2tog* is a common right-slanting decrease. The different decreases make the boot more symmetrical-looking, neat, and attractive.

Row 5. k11, ssk, k1 (c.t.s.), k2tog, k12.

Row 6. k10, ssk, k1 (c.t.s.), k2tog, k11.

Row 7. k9, ssk, k1 (c.t.s.), k2tog, k10.

Row 8. k8, ssk, k1 (c.t.s.), k2tog, k9.

Row 9. k7, ssk, k1 (c.t.s.), k2tog, k8.

Row 10. k6, ssk, k1 (c.t.s.), k2tog, k7.

Row 11. Increase 1, k5, ssk, k1 (c.t.s.), k2tog, k5, increase 1, k1.

Row 12. k5, ssk, k1 (c.t.s.), k2tog, k6. The shoe is now completely covered, and the body of the boot has begun to take shape.

NOTE: Now is a good time to tie in the cast-on ends, by weaving them in and out of the drilled holes. The other two ends can be weaved into the purl side of the fabric, as in normal knitting.

6. FILL IN THE FRONT

6a. Face the front toe area and put 3 of the centermost stitches on a DPN.

6b. Cut the yarn at the back of the shoe (leaving a tail) and use that ball of yarn to knit the front 3 stitches on the DPN.

6c. Turn the shoe so the back heel faces your body. Transfer 1 open stitch (held by the other DPNs) to each side of the DPN holding those 3 centermost stitches.

6d. You now have 5 stitches on the centermost front DPN. Purl those 5 stitches.

NOTE: This is a good time to weave in any extra ends or tighten up any holes or loose spots in the knitting, since we are now going to tightly enclose the shoe.

6e. Turn the work again to face the front toe area, and transfer 1 open stitch to each side of the same centermost DPN.

6f. You now have 7 stitches on the centermost front DPN. Knit those 7 stitches. You should now have 14 stitches on all the DPNs around the shoe.

6g. Facing the back heel, add 2 more open stitches to each side of the centermost DPN.

NOTE: At this point, you may have to break up the center stitches with more DPNs.

6h. You now have 9 stitches on the centermost front DPN. Purl those 9 stitches.

6i. Turn the shoe to face the front toe area and add 2 more open stitches, as before.

6j. You now have 11 stitches on the centermost front DPN. Knit those 11 stitches.

7. GROW THE BOOT

7a. From where you left off, start knitting around for 10 rounds.

NOTE: You may knit around for as many rounds as you want, depending on how much yarn you have left. But remember to leave enough yarn for a final rib cuff and the bind-off.

8. RIB AND BIND OFF

8a. After your last round is completed, begin a knit purl rib pattern (k1p1) for 5 rounds.

8b. Bind off all 14 stitches loosely and tie in any remaining ends.

Repeat steps 2-8 for the second boot.

Voilà! You've just made yourself a rustic pair of knitted slouch boots.

FINISH ⊠

ANT FARM
ROOM DIVIDER
By Matt Maranian

DIVIDE AND ENHANCE
YOUR SPACE WITH CARPENTRY
AND COMPUTER SKILLS

▶▶ Those tired rice-paper shoji screens aren't your only option for creating division within an open space. Inspired by the classic tabletop ant farm, this design is nearly all the fun of the living variety, plus you can sleep nights knowing there won't be any escapees.

For armchair entomologists this divider offers a seamless segue to share ant trivia with friends: Did you know that ants yawn? They can lift 20 times their own weight. Some steal pupae from other nests, hatch them, and raise their captives as lifelong slaves!

The process for this project might seem a bit involved, but keep in mind you're really only working with five materials: wood, paint, screws, plexiglass, and curtain rods. Simple, really.

» Screens are a way of creating mystery, like a carnival mask.
en.wikipedia.org/wiki/Mask

» Long-running 60s TV show *Petticoat Junction* opens with a classic strip-tease.
imdb.com/title/tt0056780

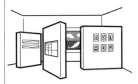
» Most contemporary art galleries use moveable walls to expand and contract space as needed for exhibition.
craftzine.com/go/gallery

» The Berlin Wall, an iconic symbol of the Cold War, came down in 1989.
en.wikipedia.org/wiki/Berlin_Wall

Matt Maranian is a designer and bestselling author whose books include *PAD* and *PAD Parties* (Chronicle Books). A long-time Angeleno, he fled the LA Basin in 1999 for the hills of Brattleboro, Vt., where he owns a new and vintage clothing store with his wife Loretta.

Illustrations by Tim Lillis

WHAT YOU'LL NEED

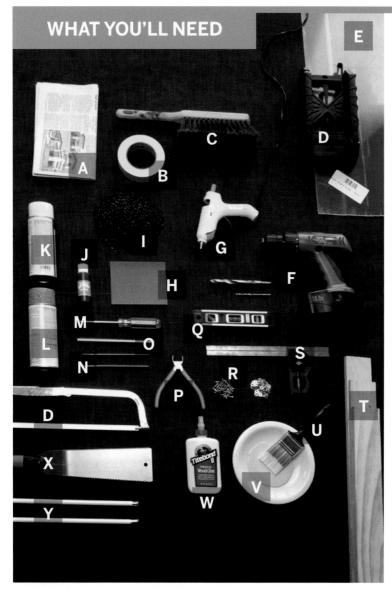

[J] Black acrylic paint

[K] White spray paint
Formulated for plastic.

[L] Gray faux-granite finish
spray paint

[M] Phillips screwdriver

[N] Fine-tipped paintbrush

[O] Pencil

[P] Wire cutters

[Q] Level

[R] #6 ½" pan head
sheet metal screws
with washers (32)

[S] Square

[T] 1"×3"×6' pine planks (3)
Look for flat, straight,
knothole-free boards.

[U] 2" wide paintbrush
To paint the framework.

[V] Flat latex paint
For the framework.

[W] Wood glue

[X] Carbon steel handsaw
or any fine saw to cut metal

MATERIALS AND TOOLS

[A] Newspaper to cover
work surface for painting.

[B] 1½" wide masking tape

[C] Large, dry paintbrush
or soft whiskbroom

[D] Miter box and saw

[E] 36"×54" sheets of
1/10" plexiglass (2)
Do not remove the
protective plastic.

[F] Electric drill with ½"
and 3/16" bits

[G] Hot glue gun and
glue sticks

[H] Medium-grit
sandpaper

[I] 5 lb. bag of
aquarium gravel

[Y] 7/16", 28-48" spring
tension curtain rods (2)

Computer, scanner, printer
and paper, 2½" drywall
screws (4), clamps, safety
goggles, soft towel and
glass cleaner (not shown)

Photograph by Meiko Arquillos

▶▶ MAKE YOUR OWN
BUGGED-OUT ROOM DIVIDER

Time: A Weekend   Complexity: Medium

1. BUILD THE FRAME

1a. Using the miter box to make a 45 degree angle, cut 2 of the pine planks at 4½' lengths, and 2 at 3' lengths (measurements given are from the longest mitered ends). Lightly sand corners and edges of the cut ends.

1b. From the tip of a ½" drill bit, use masking tape to mark a length of 2". With a pencil, mark points on the long side of each of the 3' sections, centered at a point 8" from each end of the narrow edge. Drill a hole at each of the marked points, using the 2" marking on the drill bit as your stop guide. Brush away sawdust from each piece.

1c. Clamp 1 long frame board to a table, and then clamp the short frame board flush with 1 end, clamping across the joint and making sure that the corner is square. Drill a ⅛" hole through the edge of the short board and into the mitered end of the long board. Unclamp the corner and apply glue to both faces, then re-clamp and screw the corner together using 2½" drywall screws. Wipe off excess glue, repeat for the other 3 corners, and let the whole thing dry.

2. DESIGN THE TUNNELS

2a. Remove 1 side of the protective plastic sheeting from 1 piece of plexiglass. Lay the plexi on the floor, exposed side up, over newspaper.

2b. Start by creating a horizon line, about 14" from the top edge of the 3' end of the plexi. To mask the horizon line and tunnels, tear the masking tape in half lengthwise, and working in sections, use only the torn edge for the template lines.

2c. From the horizon line, create your network of tunnels, roughly 2" in width, by either using this design as your guide, or creating your own.

2d. Tape off a 2½" border along the 3 edges of the plexi below your horizon line.

NOTE: To cut plexiglass, score a deep line, using a straightedge. Sandwich the plexiglass between a straight board (like a 2×4) and a table with the scored line about ½" away from the edge of the table, and clamp it in place. Put on your safety goggles. With even pressure and a quick downward motion, snap the plexiglass along the line.

3. POUR IN THE "SAND"

3a. With newspaper, cover everything above the horizon line. In a well-ventilated area suitable for spray painting, place the sheet of plexi — masking-tape side up — over the newspaper. Loosely sprinkle the aquarium gravel over the face of the plexiglass, making sure the gravel doesn't clump or cover solidly.

3b. With the granite finish spray paint, spray the entire prepped area, coating evenly and lightly (better to do 2 light coats than 1 heavy coat). Dry completely, for at least 4 hours.

3c. Carefully lift the plexi upright, allowing the gravel to fall to the newspaper. Use a dry paintbrush or whisk broom (or your hand) to break loose any gravel that adhered to the paint. Collect the gravel and reserve.

3d. Brush the painted surface clean, and using the white spray paint, evenly coat the entire gravel-treated surface of the plexiglass to an even opacity. Dry completely, then remove the masking tape.

4. FINISH THE FARM

4a. Lay painted side down, onto the floor or a large work surface. Take the second piece of plexiglass, remove the protective plastic sheeting from both sides (or only one, if the sheeting is clear), and place squarely on top of the first. Lightly tape the corners in place to keep pieces from moving.

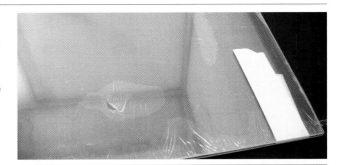

4b. Again, mask off the template for the horizon line, the network of tunnels, and the border, but this time using your first pattern as your guide, tracing through the plexi.

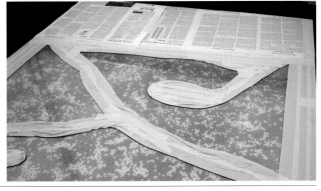

4c. Once the template is complete, separate the plexiglass. With the second one, place newspaper above the horizon line. Cover the unmasked side with the reserved protective sheeting. Repeat the same gravel/painting process as before.

5. ADD THE ANTS

5a. Once dry, scan the ant silhouettes provided here to a maximum height just short of 2" (144 pixels). Print. To get more ant action, flip the scans horizontally and print again, using mirrored images to give your ants more variety.

5b. Lay a plexi sheet over the newspaper, painted side up, and position the printed ants underneath it, inside the tunnels and along the horizon. (Don't overcrowd — it looks nicer if you allow ample space between ants.) With the fine-tipped paintbrush, trace-paint over the ant templates using black acrylic paint. Dry.

5c. Repeat the process on the second sheet of plexi, placing ants in the vacant spaces left on the first sheet. The purpose here is to give your ants three-dimensional appeal, without overlapping ant images on top of one another. Allow to dry.

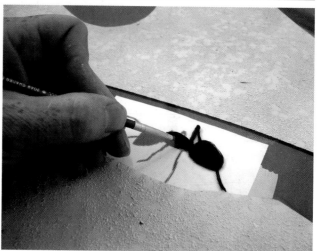

6. PREP THE DIVIDER'S LEGS

6a. There are 2 pieces to the curtain rod: a narrow end that slides into a larger rod. Unscrew the narrow rod from the larger one and separate. Unscrew the spring from the narrow rod and remove.

6b. Measuring from the rubber-tipped end, cut larger rods down to 21" with a fine carbon steel saw (your remaining piece will be around 5½" long, bearing 1 cut end and 1 finished end). With wire cutters, cut each spring to 14".

6c. The narrow rods will have 1 rubber-tipped end and 1 unfinished end. Note the dimple stamped into the unfinished end. This is used to guide the rod along the spring for adjustment, and we need to keep it. Measuring from this end, cut each rod to 14". Then take the rubber foot off of the finished end and put it onto the end you just cut.

7. ATTACH THE LEGS

7a. Lay the framework on the floor. Working on 1 end of the framework, fill 1 of the drilled holes with hot glue, and insert the 21" rod. While the glue is setting, use a square and a level to establish a straight positioning of the leg. Repeat with the other 21" rod. This will be the bottom end of the divider.

7b. Working on the opposite end of the framework, repeat the process with the 5½" lengths of the remaining larger rods, inserting the cut edges.

7c. Once the glue has set, screw each spring, cut end first, into each of the 14" narrow rods, leaving 2" of the spring exposed. Firmly slide each of these rods into the top anchor rods. From here, you will be able to screw down or unscrew the narrow rod — just as you would a curtain tension rod — to adjust and secure the divider to your desired ceiling height.

8. SECURE FARM TO FRAMEWORK

8a. Lay one of the plexi panels, painted side down, over the framework and center it. Make marks along the edge of the plexi at 9" intervals, centered over the frame. These marks are where you will drill holes to attach the panels to the frame. Take the panel off of the frame and drill a ³⁄₁₆" hole in the panel at each mark. Be careful not to press too hard while drilling to avoid cracking the plexiglass. Repeat with second panel.

8b. Lightly wipe away dust and fingerprints from the painted sides of the plexi with a soft towel and glass cleaner, being careful not to affect painted areas.

8c. Lay the framework over newspaper and secure each side of the plexiglass; space the screws/washers at 9" along the top and sides of the framework. Screw the panels onto the frame using the pan head screws and washers, making sure the panels are centered on the framework.

NOTE: This design has been sized to accommodate a maximum ceiling height of 7½', but can be adjusted to fit a higher or lower ceiling.

8d. Now set up your divider in a spot where you can marvel at your farm while enjoying a new atmosphere.

FINISH

CATNIP CASTLE
By Julia Szabo

A KITTY PLAYGROUND
WE CAN ALL ENJOY

▶▶ As a pet lifestyle expert, I'm often asked the question, "What do cats want?" The answer is simple: A fun outlet for their instinctive needs to scratch and climb.

Unfortunately, most commercial pet playgrounds consist of towers (aka "cat condos") and scratching posts swathed in hideous nylon broadloom.

If you dread displaying these eye-sores at home, you can now make your very own Catnip Castle. Easy to craft out of corrugated cardboard, it mounts to the wall, niftily saving room in small spaces. While the Castle's sculptural lines are aesthetically appealing, it also works as furniture insurance, keeping Kitty's claws gainfully occupied.

» Windsor Castle in England is one of the oldest and largest occupied castles in the world.
craftzine.com/go/windsor

» The Chinese language is derived from pictographic line drawings representing a certain object's shape or characteristic. The Chinese character for mountain looks just like a mountain.
craftzine.com/go/chinese

» One of the most famous cats in TV history is Morris, the mascot for 9Lives cat food.
craftzine.com/go/morris

» In Burley, Idaho, a DHL delivery driver reported being chased by a mountain lion while attempting to leave a package at a home.
craftzine.com/go/lion

Illustrations by Tim Lillis

Julia Szabo is the author of *Animal House Style: Designing a Home to Share with Your Pets* (Bulfinch Press). To read more about her work with animals and design, visit animalhousestyle.com.

WHAT YOU'LL NEED

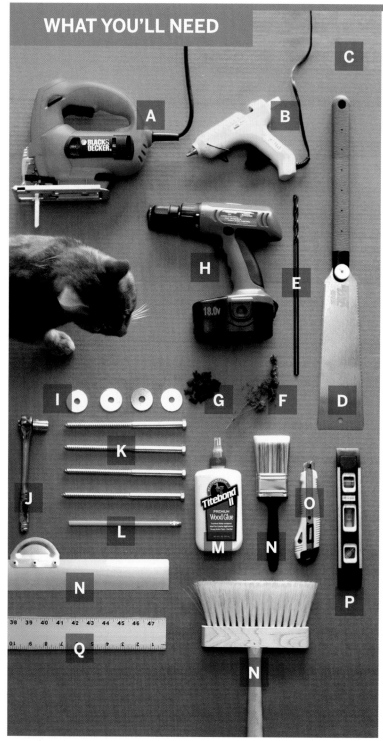

MATERIALS AND TOOLS

[A] Jigsaw

[B] Glue gun

[C] Corrugated cardboard pads measuring 36"×48" Available in bundles of 50 from uline.com. You only need 35 boards, but it's good to have extra so you can select the most perfect ones.

[D] Hand saw or reciprocating saw

[E] ⅜" and ⁹⁄₃₂" wood drill bits, 6" or longer

[F] Dried, loose catnip

[G] Cat treats Liv-a-Littles cod treats (halopets.com) work exceptionally well.

[H] Power drill

[I] ⅜" washers, 2" diameter (4)

[J] Ratchet with ½" socket

[K] 8"×⅜" lag bolts (4)

[L] No. 2 pencil

[M] Carpenter's glue

[N] Paint brush, wallpaper brush, and trim guide for spreading glue; available at paint supply stores.

[O] Sharp utility/camping knife I prefer Spyderco.

[P] Level

[Q] 48" rule

Heavy objects to weigh down glued boards I used a large piece of plywood, plus several unopened bags of plaster (not shown).

A powerful vacuum cleaner I like the Dyson Animal (not shown).

Photograph by Meiko Arquillos

⏩ TURN CORRUGATED CARDBOARD INTO A SCULPTURED CAT PAD

Time: A Weekend Complexity: Easy

1. GLUE CARDBOARD TOGETHER

1a. Pour carpenter's glue generously onto the center of the first board, then spread the glue out with the brush and trim guide. Take care not to pour it on too thick near the edges, or the glue will seep out, leaving unsightly ooze that dries yellow. Lay the next board on top of the glued board, taking care to match up the corners, and repeat until you've attached 10 cardboard pads together. Repeat this process with the next 10 pads. Finally, glue 15 pads together the same way. You now have 2 cardboard blocks of 10-pad thickness and 1 of 15-pad thickness.

1b. Carefully place something heavy and flat over each stack, such as a piece of plywood, taking care not to shift the boards, and leave to dry and adhere for several hours, or even overnight.

2. CREATE A DESIGN

I explain how to create an abstract castle featuring a tall central tower flanked by a battlement on either side. Of course design possibilities are limitless, and whether you want to duplicate my design or create your own is up to you.

2a. Take one of the stacked pads and, following the diagram shown here, start at one 48" edge and mark the following points with your rule: 6", 21", 27", and 42". Then mark 6" up from the bottom at all 4 points, and rule the lines down to those points. Repeat with the other 2 stacks.

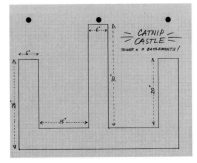

NOTE: If you're designing your own cat pad, plan ahead and draw a diagram on graph paper before beginning the project.

3. SAW AND ADHERE

3a. Saw carefully along the lines drawn. Repeat until all 3 cardboard stacks have the Castle shape.

To create the "battlements" on either side of the central "tower," rule a line 10" down from the top on each, and saw carefully along the lines. Remember to keep your jig saw moving; if you need to stop in the middle of a line, do not turn off the saw; keep it running or you'll produce a raggedy effect.

When you saw through the section that's 15 pads thick, you will need to complete the sawing job with a handsaw (I used a pull saw) or reciprocating saw with a long enough blade.

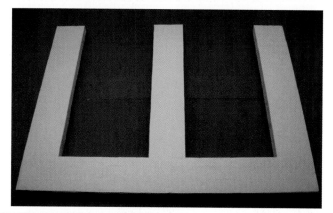

3b. Check the edges of each section carefully. You want them to be securely glued. If they're not, they will come apart and fan out a bit at the edges. If this happens, carefully use a glue gun to insert hot glue between the boards, then press them together again.

3c. Glue together the 3 sections comprising the Castle, so that you now have a stack of 35 pads. Stacked and glued together, 35 boards will measure approximately 5¾" deep.

As before, weigh them down until they are dry, carefully keeping the edges as flush as possible. Repeat with the 3 sections that comprise the stepping blocks.

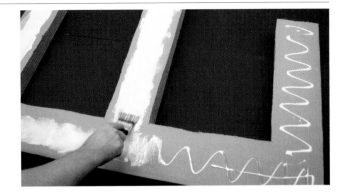

Q: I've cut out the Castle and glued the 3 stacks together, and now I notice a few uneven edges. How can I smooth them out?

A: You can smooth down any offending imperfections with a jigsaw, or a reciprocating saw with a blade at least 6" long. Before using the jigsaw to carve out the rectangular steps (15"×30") that lead up to the Castle, first make a pilot cut pointing straight down at each corner with your utility/camping knife to ensure the cleanest possible corners.

4. SIZE TO FIT

4a. Customize the design so that it works most efficiently in your space. The number of corrugated stepping blocks you will need to provide for Kitty to reach the Castle depends on the height of your ceiling. If it's low, you'll need only 1 on each side; higher ceilings will require 2 on each side.

Because my wall is only about 8' high, I needed only 1 step on either side of the Castle for Kitty to reach her goal, so I divided one of my 15"×30" blocks equally into 3 sections measuring 10"×15" each.

If you have a taller ceiling, you can use the 10" long pieces you sawed off to create the battlements as 2 additional, smaller steps, or saw the third 10"×15" section in half. If you'd prefer 2 taller steps, saw the second 15"×30" block in half lengthwise for two 7½"×30" steps, and shorten by sawing as desired.

Alternatively, you can also attach your Castle to a wall above a piece of furniture, such as a sofa or chest of drawers. With that configuration, Kitty can use the furniture as a springboard to reach the Castle.

5. INSTALL

5a. To determine your Castle location, find the studs in your wall. An electronic stud detector is a wonderful thing, and if you know someone who has one, by all means borrow it. I used the more primitive divining method of knocking at the walls.

NOTE: You'll need different supplies depending on the type of wall. If your wall has aluminum studs (as ours did), you'll need only the supplies listed, plus a wood bit. If you're working with a brick-and-concrete wall, you'll need to drill holes with a masonry bit, then use a hammer to insert lag shields measuring ⅜"×1¾", to serve as anchors for the lag bolts.

5b. Use a pencil to mark the wall where the Castle's top edge will go, and use a level (this is a large item, and if it's not level, it will look sadly amateurish). It's important that the Castle be attached solidly to the wall so it doesn't wobble; if it moves when Kitty first jumps on it, she won't feel safe jumping on it again.

For petite Mademoiselle, I centered 2 bolts on the central tower plus one for each stepping block. (For a heavier cat or multiple cats, 2 bolts per stepping block are recommended, plus one centered on each battlement.)

5c. Use a pencil to mark the first drill hole 3" up from the bottom of the tower, and the second one 10" from the tower top, aligned with the top of the battlements.

For the stepping blocks, avoid placing your drill hole at the center, or the block is liable to spin around and deter Kitty. I centered it 3" down from the top (narrow) edge.

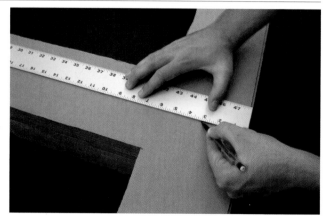

5d. Using a ⅜" drill bit long enough to go through the 6" of cardboard, drill through your pencil markings all the way to the other side.

5e. Have a friend help you hold the Castle up to the wall, lining it up with the marks you made for the top edge, and insert a pencil into each hole to mark where you'll be drilling on the wall (this helps ensure accuracy). Then lay the Castle back down on the floor and drill 2" into the wall at each marked point, with the ⁹/₃₂" drill bit.

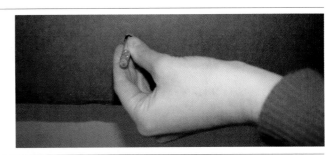

5f. Put each lag bolt and washer together, then insert into the drill holes. Using a rachet, screw into the wall until snug.

NOTE: Tighten until the cardboard surrounding the bolts is compressed slightly, as shown (you don't want them too tight or too loose).

6. SAY "HERE, KITTY KITTY!"

6a. To attract Kitty to her new playground, rub it thoroughly with organic dried catnip. Crush the catnip between your fingers to release the aroma and make it extra fragrant. Then rub catnip all over the sawn surfaces, using extra for the sides of the central tower to lure her up there. If Kitty isn't compelled to make the initial leap, insert small pieces of her favorite dry treat in the corrugate openings — that'll get her moving.

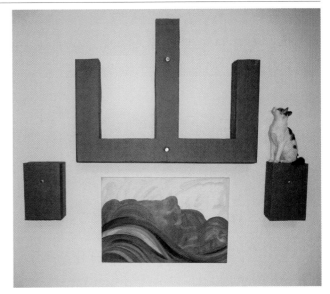

So, how does the Castle rate with the toughest customer, a domestic short-hair cat? Mademoiselle, my pastel calico model, wouldn't say, but her reaction spoke volumes. As soon as the Castle was installed, she promptly got busy connecting with her inner tiger, navigating the vertical maze, scratching, stretching, leaping, pausing calmly atop a battlement to groom herself and purr with satisfaction, and generally regarding the Catnip Castle as her royal domain.

FINISH X

BAZAAR

CRAFTY GOODS WE ADORE. *By Natalie Zee*

Reprodepot Fabrics

reprodepotfabrics.com

● Featuring a wide selection of stylish, vintage, and kitschy prints, Reprodepot Fabrics can easily be declared the best online fabric store. A recent fabric purchase included Parisian Style (a Japanese import fabric with drawings of manga-eyed French women), a whimsical, brightly colored polka dot print called Light Bright, and my favorite, Fashion Kitty, featuring a 1950s-style kitty in fun poses.

Kitsch to mod, Reprodepot delivers the fabbest fabrics, online.

Subversive Cross Stitch

33 Designs for Your Surly Side, by Julie Jackson $15
chroniclebooks.com

Subversive Cross Stitch can also double as good crafting therapy. Almost anyone can pick up this art form and create their own messages filled with wit, sarcasm, and pure inspiration. My favorites, "No You DIDN'T!" and the homage to 80s TV waitress Flo's signature line, "Kiss My Grits," are at the top of my projects list.

Kiku Knitting Needles

$10
kiku-co.com/stitched.htm#knittingneedles

● Here's a surefire way to get noticed the next time you meet up with your knitting circle. Kiku's knitting needles are handmade and topped off with pure panache. Choose from devil duckies, tikis, happy Buddhas, pirates, skulls, fez monkeys, and sushi. All needles are 12" long and come in sizes 7 US, 10.5 US, and 11 US. Makes a great gift for your knitting friends!

Stitch Lounge

San Francisco
stitchlounge.com

● Stitch Lounge was started by three lifelong friends, bonded by their love of crafting. Their concept was simple: create a modern-day sewing circle in a city where large sewing space is typically sparse (i.e. your sewing machine is stored in your closet).

The studio boasts a large cutting table, a number of industrial-strength sewing machines, sergers, and other sewing notions and equipment. You can drop in and pay for studio time, or take a class on a variety of topics such as sewing basics, handbags, hems, and basic alterations, or refashioning classes like "T-Shirt Reconstruction!" I walked in with a bunch of MAKE T-shirts and left two hours later with a cute flutter-sleeve top.

If you don't live in the Bay Area, don't fret! The dynamic trio has come out with a new book, *Sew Subversive* (Taunton Press), which is filled with lots of inspiring DIY fashion, refashioning ideas, and tutorials for projects like making a pillowcase dress, or turning old sweaters into a scarf.

The stylish refashioned MAKE T-shirt.

« Xyron 900

$100

xyron.com/enUS/Products/Xyron_900.html

The Xyron 900 is a fun machine that makes stickers, magnets, labels, or laminate. It's large enough for letter-sized paper, yet small enough to fit on a desk. I make birthday cards that look like cheesy tabloids starring my friends. I design in Photoshop, print, then use the Xyron to make a sticker, which I then mount onto thick cardstock.

Crafting gadgets are cool!

« Denise Interchangeable Knitting Needles

$50

knitdenise.com

I have lots of knitting needles, but I'm never able to keep track of what I have. I end up buying a new pair with every new project in fear of not having the right size when I get home.

Luckily, I came across the ultimate knitter's toolbox. Denise Interchangeable Knitting Needles is a wonder kit that contains everything you need to create needles in ten different sizes, circular or straight, and in variable lengths. Plus, the box is super compact for storage.

This kit is the best investment for any knitter's toolbox.

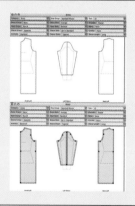

« Garment Designer 2.1

$165

cochenille.com/garm.html

Garment Designer is a cross-platform application that lets you create your own sewing or knitting patterns for tops, skirts, pants, and dresses. It offers pre-existing patterns that you can customize with easy drop-down menus and size fields. Just print out the patterns and go!

« The Sampler

$23/month, $53/3 months

homeofthesampler.com

Subscribe to the Sampler and you'll receive a monthly collection of samples and promo goodies from indie crafters, record labels, and zines. When I got my first Sampler package, I was surprised at how much stuff it contained. Some of the cool things I got: a knit sunglasses cozy by Magic Yam, a "pencil for writing love letters" from ohbara.com, funky earrings from zoo-loo, a mini newspaper edition of *The Crafty Times*, stationery, cards, and buttons galore.

CRAFT LOOKS AT BOOKS & KITS

« The Sunlight Print Kit
By Paul Grivel, ISBN: 0811852636 $23

chroniclebooks.com

Sunlight prints are photographs made from special light-sensitive paper and, of course, the sun. No chemicals are needed to develop the photos, just a little bit of lemon juice. This kit includes everything you need to make sunlight prints, as well as a booklet with plenty of great photography techniques and how-tos.

« Get Crafty's Guide to Hip Home Economics
By Jean Railla, ISBN:978-0-7679-1720-9 $15

randomhouse.com

If you're new to crafting, this book will help you find your inner crafter. Author Jean Railla, founder of getcrafty.com, instructs you on every-day tasks ranging from making your own household cleaners or face scrub to making vinyl-record flowerpots and sewing an A-line skirt.

« The Craftster Guide to Nifty, Thrifty, and Kitschy Crafts
Fifty Fabulous Projects from the Fifties and Sixties
By Leah Kramer, ISBN: 1-58008-747-7 $18

tenspeedpress.com

Vintage crafts from the 50s and 60s take center stage thanks to Leah Kramer, founder of hipster craft site craftster.org. Talk about retro goodness! The "Painted Glass Martini Set," "Powder Room Poodle," and "Temptress Collar" are just a few of the projects that will get your pad swinging and your closet all dolled up.

« Paper Craft Kit
Materials and Instructions for Beautiful Handmade Paper Creations
By Sidd Murray-Clark, ISBN: 0811848280 $25

chroniclebooks.com

I collect all kinds of paper and love everything paper-related. For all you paper lovers out there, this kit not only comes with 20 sheets of beautifully colored and patterned paper, but it's also filled with fun paper projects. Complete with dowels, stencils, and a scoring tool, you'll be able to make a wide array of projects, including a translucent paper lantern, flying kite, and concertina book.

Natalie Zee is an associate editor of CRAFT and reports on all things crafty at craftzine.com. Her love of fashion can also be seen on her personal blog, Coquette, at coquette.blogs.com.

101: SILK-SCREENING

By Kirk von Rohr

Print your designs on anything you can hang, wear, or tote.

Silk-screening is a great way to personalize your gear. It's a very basic process that has unlimited outcomes. One of the easiest ways to get a design on almost any surface is to use the photo emulsion process. Once you've made the screen, it's ready to print time and time again. Follow along as I walk my colleague Sara Huston through the process of transferring our design to a screen, and printing it on a laptop bag.

> ✳ **LOCATION:** You can do this at home. All you need is a sink and an open workspace. Special thanks to Ape Do Good for letting us shoot this project at their screen printing studio in San Francisco. Check them out at apedogood.com.

GATHER »

Work up an idea for your design. On your first attempt, try a one-color design, keep it simple, and have some fun with it. Once you get it figured out, make your design digital. Sara and I created ours in Illustrator, but you can also scan a drawing. If you are really hands-on, you can draw straight onto transparency paper using India ink. You need a solid black positive to burn into the screen. I print on transparencies, using a black and white laser printer. This gives me an easy way to accurately, cheaply, and quickly create a positive.

SARA: Are silk-screening inks special? How do they differ from other paints?

KIRK: Silk-screening inks are acrylic. They dry quickly, and are water-soluble and transparent. Due to the transparent nature of the ink, it will interact with the color of the object you are printing on (example: blue ink on yellow shirt will turn slightly greener; blue ink on a black shirt will be barely legible).

White, black, and a few metallics are the only completely opaque ink colors. White also comes in extra opaque, which I recommend. If you want to print on a dark material, mixing some opaque white into the color will help it stand out.

MATERIALS

» DIAZO PHOTO EMULSION MADE BY SPEEDBALL
» 8"x10" SILK SCREEN
» 8"x10" PIECE OF GLASS
» SQUEEGEE
» TASK LIGHTS (2)
» 150-WATT BULBS (2)
» TRANSPARENCY PAPER FOR BLACK AND WHITE COPIER/LASER PRINTER
» SILK-SCREEN INKS (CREATEX AND SPEEDBALL HAVE WORKED WELL FOR ME)
» LID TO IKEA BIN FOR STRETCHING SHIRT OVER, OR OTHER HARD FLAT SURFACE YOU CAN SLIP INTO A SHIRT

OPTIONAL
» FAN (I USE A SMALL VORNADO)
» DIAZO PHOTO EMULSION REMOVER (IF YOU WANT TO CLEAN YOUR SCREEN AND START OVER)

DESIGN »

Design must be high-contrast. There is either on or off, positive or negative. Grays are acquired by decreasing the size of solid dots. This can be done using halftone dots.

Take a bold approach. Bolder lines print more easily. Save your delicate designs for when you get a feel for the process.

Mind your solids. Be cautious of big solid areas. They are harder to print because they require lots of ink and even ink coverage. On fabrics, they also get rubbery, and will eventually crack after lots of washing.

Have fun with it.

✳ **MORE:** Go to craftzine.com/01/101 for more designs to print.

1. PREP THE SCREEN

Mix the photo emulsion as per the directions.

Coat the screen with photo emulsion, working fairly quickly over a sink or surface you can get messy. It's OK to have indoor lights on during this process, but keep out of direct sunlight. The emulsion needs to be applied evenly, so keep flipping the screen over and squeegeeing until the emulsion is even on both sides. Any globs will cause uneven exposing and will mess up your end result. The thicker the emulsion is applied, the longer the screen will have to be exposed.

The screen needs to be completely dry in order to expose it, and should be dried in a pitch-black room. I dry my screen by resting the wood frame on a couple of shoe boxes in the closet, so that the screen is parallel to and above the floor. This allows the air to flow above and below the screen to help it dry faster. Make sure that only the frame touches the

 SARA: You can really see the fibers in this silk — it looks really open — does that help the printing process?

KIRK: The tighter the weave of the silk, the better resolution you will get. Think of it like dpi (dpi = dots per inch). If your design has lots of detail and you make it 72dpi in Photoshop, of course you'll lose detail and it'll look grainy. If you go to 150dpi, it will look twice as good. If your silk is cheap, chances are it's less than 72dpi.

To gain resolution, you will have to buy more expensive silk, and in most cases, that means stretching your own screen. I go the quick and dirty route and use store-bought pre-stretched screens, which cost around $10 each for an 8x10. In using these, though, I know that any wispy details in my design may be compromised.

boxes, so as not to mess up the nicely applied emulsion. You can place a fan (I use Vornado because they are compact) next to the screen. Drying it this way takes 30 minutes to an hour, depending on humidity.

2. EXPOSE THE SCREEN

Now that it is dry, place the screen on your workspace with the bottom facing down. Put your transparency on the screen in the center and as squarely as you can, then place a piece of glass on top. This holds your transparency down so that it makes direct and even contact with the screen. If it doesn't make direct contact, then your design will appear fuzzy around the edges.

The light source needs to be placed about 12 inches from the screen to get good results, and it needs to shine evenly across your design. I use two $10 task lights. These are great because they allow me to easily adjust my light source, and by having two, one on either side of the screen, we can make sure the entire design gets an even, direct supply of light. Follow the directions that came with the emulsion for exposing your screen. It varies with the bulb and screen size. I'll burn our screen for about 30 minutes. You can tell when the screen is done by looking: the exposed areas turn dark green when they are baked solid by the light.

 TIP: For a super-dense positive, make two trans-parencies with your design on them. Line them up and attach them together with double-sided tape.

SARA: Which way does my design go? Does it need to be upside down or backwards?

KIRK: Nope, what you see is what you get. You should lay down your design onto the screen just as you want it to print. If you have type in your design, it should be right reading.

3. WASH AND DRY SCREEN

Now that the screen is exposed, wash it off in the sink with hot water. It takes some force to wash the screen effectively. I've attached a special nozzle to my faucet that creates higher pressure. (I got a nozzle at Bed Bath & Beyond for $5. Just screw it on and it'll toggle between high and low. Works great for dishes too; I leave it on all the time.)

Along with spraying, you can gently rub the screen with your fingers. Don't use your fingernails. If you force the emulsion off, you run the risk of tearing off the hardened emulsion, putting you back to step 1. You want only the unexposed area to wash off. Under hot water, the emulsion will become slightly gummy. Drying the screen isn't such a big deal this time around, now that it isn't sensitive to light. Prop it up against the fan, or place it where it can get some air. Silk dries quickly.

4. PRINT IT

Now that the screen is exposed, washed, and dried, print it and see how it works. Try it out on paper first.

Lay the screen down flat, making sure that your surface is even and flat.

With a spoon, put a glob of paint on the screen and spread it the width of your design. Don't get any on the design itself, just the area above it.

Now the fun part. Hold the screen down firmly with one hand (or have a buddy help hold it). Use a squeegee to pull the ink down to the bottom of the screen. Apply a small amount of pressure to the squeegee as you pull the ink. You will be able to see the paint evenly distributed across the screen.

Lift the screen and look at your beautiful print! Be very careful when you lift off the screen. Try to peel

 SARA: How can I be sure that my print will be even?

KIRK: The word on the street is that you should only have to spread the paint downward once, but if you don't think it's looking even, give it another pass. With practice, you'll get a feel for it.

SARA: I can feel the squeegee sticking in certain spots; why is that?

KIRK: When you pull down, you should feel it evenly sliding over the screen. If it's grabbing in areas, then there isn't enough ink in that spot. This is one of the telltale signs that your print needs another pull of ink. If the first pull feels even, it may not need a second pull.

it slowly and directly up, so you don't smudge the fresh ink. It may want to stick to the paper.

It's as easy as that! Lay the screen down on another piece of paper and do a few more prints for fun.

5. PRINT IT

Now that you have some practice and a feel for things, let's print the laptop bag. Start with a clean screen.

Since the bag is soft, we need to put something stiff inside the bag to make the printing surface a little harder. In this instance, I used my old cutting mat.

Put some masking tape down on the bag as a guide to help line up your screen. The emulsion is just slightly transparent, and you can see the tape through it. Once it is into position, hold it down, glop some paint, and make a nice swish of the squeegee. Lift off the screen and take a look. Beautiful!

✳ **TIP: Do a few test runs so you can practice getting good ink coverage and squeegee pressure. It's good practice to test your screen on some scraps of material that are similar to what you want to put your design on.**

6. CLEAN UP

Place your finished product somewhere to dry (it will take 15 to 30 minutes). Immediately wash your screen — the ink dries fast and can ruin your screen.

You can make about 100-200 prints with your screen. When doing a long run, you may have to periodically wash out your screen between prints to keep the paint from clogging your design.

MAKE IT WASHABLE

If you are printing on a shirt, you need to do another step to make it permanent. Here are two options:
» Iron the shirt on a high, dry setting, placing wax paper between the shirt and the iron.
» Add a few drops of an additive like Versatex Fixer to the paint before you apply it to the shirt. You can mix it right into the ink container.

SARA: I was always a bit hesitant to try silk-screening. It always seemed like I would have to take a semester-long class to learn to print a T-shirt. Kirk's process is something that I can actually do at home, and now I'm not scared to get a little magenta paint under my fingernails, instead of just on top.

KIRK: It really is pretty simple. And once you get the hang of things, you can have a screen ready for printing in an afternoon. I love the immediacy of simple replication that is inherent in silk-screening. Who doesn't love a hand-printed bag, card, or T-shirt? Now go do it!

FINISH ☒

Cell Charms: Baubles for Your Mobile

Cell charms are fun pieces of jewelry that dangle from a cellphone. This project is also a great way to use up any extra beads you have lying around, or showcase your vintage finds in a new way. This cute vintage owl cell charm, as created by jewelry designer Kris Nations (krisnations.com), takes less than 10 minutes to make!

You will need: Straight pin, large bead (like this vintage owl), 2 small round beads of contrasting color, mobile phone charm holder (available at select bead stores or online at craftzine.com/go/cellholder), pliers, wire cutter, round-nose pliers

Wire cutters   Round-nose pliers   Plain-nose pliers

1. Beading

Take the straight pin and string the beads, starting with a small round bead, so that the 2 smaller beads nest the larger one.

2. Securing

Create a small loop approximately ½" before the top bead, using your round-nose pliers to twist the wire of the straight pin. Loop the wire twice around the pin *and* top bead tightly. Clip off the excess straight pin wire with your wire cutter.

3. Connecting

Using both pliers, one in each hand, hold the jump ring of the mobile phone charm holder. Open the jump ring, with one hand twisting toward you and the other hand twisting away. Place finished charm inside the jump ring and close the ring securely.

Illustrations by Dave McMahon

Natalie Zee is an associate editor of CRAFT and writes for the craft blog at craftzine.com.

Battle Chic

Craft a wardrobe of medieval armor with DIY chainmail. BY ANNALEE NEWITZ

Photography by Quinn Norton

H enrik Olsgaard, aka Henrik of Havn, has been proclaimed King of the West six times. Obviously the guy is deft with a sword — you don't get to be King in the Society for Creative Anachronism (SCA) without winning several bouts in the annual Crown Tournament. But his triumph is also testimony to his skill at making chainmail. Henrik has been fashioning chainmail of every description for the past four decades — from beautiful, sterling silver belts to a 50-pound battle hauberk (a knee-length shirt).

And now, I'm going to teach you what he taught me: how to make your own chainmail. With just a few basic patterns, you'll have all the knowledge you need to fashion a helmet, shirt, belt, coin purse, and even a full hauberk.

Materials

» **Needlenose pliers**

» **⅛" diameter dowel**, which you can hold in place with a rack or some other device

Make rings yourself using: **⅜" diameter rod, coat hangers** (15; or several feet of stripped aluminum wire or junk wire that you want to recycle), **hammer, rod clamp** that fits the ⅜" rod, **small wire snipper**. Or you can buy pre-made steel rings from chainmail.com.

1. Buy or make your own rings.

For this project, we'll create square and triangular swatches of chainmail, which are the basic building blocks for anything larger. There are two ways to go about this: you can make your own rings from stripped aluminum electrical wire, store-bought wire, or coat hangers, or you can buy the rings ready-made from sources like chainmail.com.

If you'd like to create your own rings, begin with the wire of your choice. For lighter pieces, you might try aluminum wire. Most mail makers prefer steel. To create the rings, wind your wire around a rod, then flatten the end with a hammer and hold it in place with a small rod clamp (Figure A).

CAUTION: As you wind, you're creating a high-tension spring — if you slip, the wire can snap back at you quickly.

Next, you'll want to slip the coil off your rod and use a wire snipper to cut the rings (Figure B).

Be sure, as you're cutting, that you are creating full rings and not leaving gaps.

Shortcut: Some armor makers cut their rings out of door springs.

2. Knit your rings into chainmail swatches.

Once you have a few hundred rings, you're ready to start knitting (for reference, a full hauberk takes about 10,200 rings). To do this, you'll start by creating a small length of chain using 2 needlenose pliers to open and close each ring (Figure D) — this is the knitting part. When you pull the rings open, be sure to open them sideways (Figure C).

Your first string should be twice as long as the first row in the swatch you'll be making. For this exercise, we chose to make the string 20 rings long, for a 10-ring row. Once you have your string, you'll want to thread it onto a thin rod so that it hangs properly as you add length (Figure E). Henrik's knitting rig is typical — he's got two uprights with holes in them to place the rod into (Figure G). But you can use literally anything that will hold your rod in place as you work, including simply taping it to 2 boxes.

To create the first 2 rows, you should thread every other ring, thus creating the kind of design shown on page 115.

This is the beginning of the "four in one" pattern you'll use for the rest of your swatch. When you make your next row, you'll want each ring to link to 2 others. Each ring on the second row should have 4 rings in it.

If you want to make a square, every other row should begin with a ring that only holds one ring from the row above, as described thus far. But for a triangle, you want to shrink each row by 2. So you'll start each row with a ring that connects 2 from the row above. Keep knitting until you have a square or triangle of the size you wish.

3. Connect swatches to make armor (or anything metallic!)

Now that you're practiced, you're ready to pick a pattern and start making something other than a square or triangular swatch (see Figure H). Figure F shows how triangles and squares can fit together to create other shapes. There are several sites that offer patterns you can choose from, including theringlord.com, which sells armor supplies and pattern books and hosts a community forum, as does artofchainmail.com.

Henrik says the best part of chainmail making is getting a chance to wear and use it with friends. "I don't like getting in front of paying audiences to do this," he explains. "I like to share what I've done at SCA tournaments and places where everybody participates in the event." And that's where I find him, in full armor, a few weeks after our lesson.

Annalee Newitz (techsploitation.com) writes about geek culture in San Francisco. Her favorite dragon is black, and her favorite accessory is a bag of holding.

A

B

C

D

To make chainmail, you can buy pre-made rings or make them with stripped aluminum wire or coat hangers, which you must coil and cut. Once you have rings, you start your project by knitting square and triangular swatches. Then attach your swatches together to make metal garments and accessories such as helmets, belts, collars, coin purses, and traditional knee-length shirts called hauberks.

E

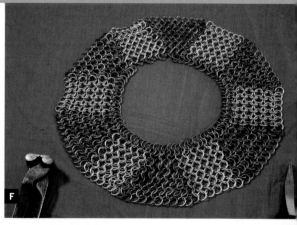

F

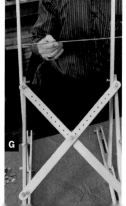

G

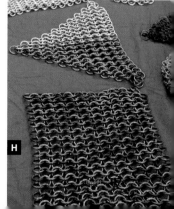

H

Wearable Artifacts
Making jewelry with everyday fossils.

BY GRETCHEN WALKER

Τhe jewelry I create is reminiscent of a more ancient time — like relics unearthed and altered with modern elements. The techniques described here let you create wearable art by texturizing metal with everyday objects found outside, at home, or in your work studio. All you need are basic soldering skills along with the materials listed at right and you're on your way to making wearable fossil art.

1. Texturize the metal sheet.
There are many ways to impart texture to a sheet of virgin copper using items you may already have, such as large and small metal washers, a metal screen (I prefer bronze); wire (iron or brass); a sharp, long nail or hand-held engraver; or a piece of concrete or sidewalk.

Place your copper sheet on a hard metal surface (metal block or anvil). Lay the objects directly onto the metal sheet, and then hammer for

Photography by Vincent Atos

Materials

» **Copper** (20 gauge sheet and 16 gauge wire)

» **Circle template**

» **Sharpie marker**

» **Metal shears**

» **Jeweler's saw**, size 2/0 saw blades

» **Beeswax** or candle wax

» **Flex shaft or Dremel tool** with 79mm drill bit

» **Hammer and metal block** or hard surface for hammering

» **Abrasives** (Scotch-Brite pads and 220-grit sandpaper)

» **Chopstick** with round shaft

» **Cotton cord** for necklace

» **Soldering block**

» **Torch**

» **Easy solder and flux**

» **Small-toothed file**

» **Flat-nosed pliers**

» **Tweezers**

» **Third hand** (flexible soldering clamp)

» **Silver black patina solution**

» **Paintbrush**

» **Baking soda**

texturized effect (Figure A, next page). You'll want to practice by creating texture on sample pieces of metal sheet. If you have the option, anneal the sheet often, because as you continue to strike, it will temper (harden) the metal.

2. Cut out the pendant.

The pendants shown here are round (although you can make them any shape you like). If you don't have a circle template, you can use the lids of vitamin or baby food containers, or even the bottom of a teacup.

Once you've chosen a form for a template, use the Sharpie marker to trace the shape onto the texturized copper sheet, then cut it out with metal shears or a jeweler's saw (Figure B). This takes a steady hand and patience.

Be aware that if this is a first attempt with the jeweler's saw, you will want to have extra blades on hand. Using your flex shaft or Dremel tool, drill a hole along the inside perimeter of the design. Place a saw blade in the top clamp of your jeweler's saw, threading the blade through the hole in the sheet (design side up). Secure the blade in the bottom clamp of the saw.

Brace the metal sheet against your bench pin by clamping your fingers against the top of the sheet and your thumb under the pin. If you do not have a bench pin, try using a brick on a tabletop.

Make sure to wax your blade, as it eases the movement through the metal — you can use beeswax or candle wax. Following the design you've traced onto the metal, take your time, and try to keep your saw vertical (side-to-side motion will break the blade).

3. Make and attach the jump ring.

Jump rings can be costly to purchase but are easy to make. First, find a chopstick that has a round shaft. Unroll 10" of the 16-gauge copper wire and anneal it to a glowing red. This will realign the molecules of the wire and make it more malleable. Take the copper wire and wrap it in a tight coil along the length of the chopstick. When the length is coiled, slide it off and begin sawing through one jump ring at a time (Figure C). Using your flat-nosed pliers, select a jump ring and close the ends tightly together. Next, using your file, create a flat surface where the two ends of the jump ring meet. The pendant will also need to be filed where the jump ring will be attached.

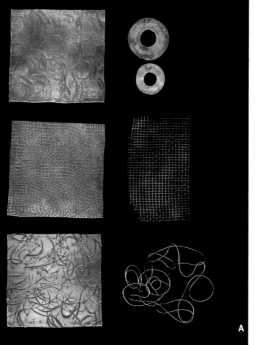

Fig. A: Lay objects directly onto the metal sheet and hammer to create these types of texturized effects.
Fig. B: Use the Sharpie to trace shapes on the copper sheet, and then cut them out with metal shears.
Fig. C: Saw through your coil, one jump ring at a time.

Set the pendant in your third hand and flux the soldering site as well as the jump ring. Solder a small piece of easy onto the jump ring and flow the solder into the cracks using your torch. Quench and reapply the flux to the jump ring. It's important to have clean surfaces, otherwise the solder will not flow easily. Use your tweezers to hold the jump ring, evenly heat both surfaces, and solder the 2 pieces together.

4. Finish the piece.

Finishing is important — it determines the overall mood of the piece. Make sure the pendant is oil- and dust-free, and then, using a paintbrush, apply a thin layer of silver-black solution to the entire surface. Silver black is an acid, so neutralize it afterwards in a bowl of water and baking soda (1 tablespoon baking soda per cup of water). Just dip it and then remove it. Using a Scotch-Brite pad or 220-grit sandpaper, buff away the black patina. Play around with the effect until you achieve the look you want. Thread the cord through the jump ring and enjoy!

Gretchen Walker is fascinated with ornamentation. She earned a degree in goldsmithing, and now lives in San Francisco, where she's both a jeweler and artist.

Quick Tip: Texturize metal with your own customized handmade stamp.

Here's how to make one:

1. Using your jeweler's saw, cut a large screw in half at the stem.

2. File down the surface.

3. Using your jeweler's saw, cut out a small star (or any small shape you want) from a 16-gauge sheet of nickel or iron (iron lasts longer but is very difficult to cut).

4. Solder the star onto the prepared surface.

No Ordinary Stitch Session

Forsake boardslides in the name of art.

BY JENNY HART

Photography by Tim Brown

E mbroidery became an obsession of mine about six years ago. Then it took a turn: I wasn't content stitching only on fabrics, I wanted to intersect new planes with stitches. Then last spring I was invited to participate in a group show of mocked-up skateboards to benefit the Knoxville Skatepark in Tennessee. This was an exhibit that would include painters Gary Baseman, Dalek, and the Art Girls. I guess I could've painted a board too, but instead, I decided to embroider it. Kickflip all you want, just go easy on the boardslide.

If you want to compromise the structural integrity of your skateboard in the name of art, just follow my easy steps! Austin-based artist Tim Brown (timlandia.net) painted some clouds on it for extra jazz.

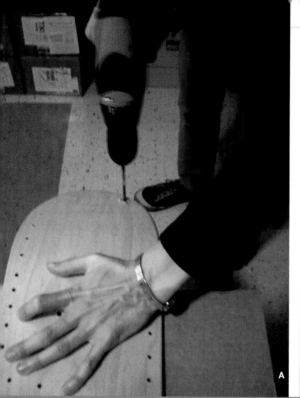

Fig. A: Each drilled hole is where a needle would intersect fabric in a traditional embroidery. Fig. B: Lay your template over the blank deck, and with an awl or tack, mark points on the board for drilling. Fig. C: Make a knot in one end of a length of leather lacing to pull through a hole and begin lacing the board.

Materials

» Blank skateboard

» Drill with a $\frac{9}{64}$" bit

» Leather lacing

» Awl or thumbtack

» Wood glue

» Tracing paper

1. Make the template.

For this first attempt, I kept the design simple. You'll have to anticipate where the thread (in this case, leather lacing) will be coming through the board and pre-drill those points.

Make a configuration of points on tracing paper based on the endpoints of an embroidery stitch (Figure B). This skateboard uses the feather stitch, whipstitch, running stitch, and straight stitch. Each drilled hole is where a needle would intersect fabric, or in this case the board.

2. Transfer the template.

Lay your template over the blank deck, and with an awl or tack, mark points on the board for drilling.

3. Drill holes.

If you intend to ride the board, you'll want to add grip tape to the top of the board before drilling. Drill holes on the points you marked (Figure A).

4. Lace the board.

Make a knot in one end of the leather lacing (Figure C). Pull through the front of deck and continue through pre-drilled holes. To make a laced edge, simply drill evenly placed holes near the edge of the deck and wrap the leather lace around the edge (see page 121).

5. Finish.

Secure knot at end and daub all knots with wood glue. Add wheels, and step on board!

Sublime stitcher Jenny Hart is known for thinking outside the hoop. Read her full Crafter profile on page 24.

Cro-bot

Crochet a robot doll that rocks. BY BETH DOHERTY

A crocheted robot — what a hilarious contradiction! Wasn't Rosey the Robot supposed to help speed things along, to save us from tedious domestic duties? But here I've spent the last week laboring over the design of a tiny, yarn-made, non-animated robot, and I would have to do some sort of voodoo spell before I could ever get it to do the dishes or mop the floor. I shouldn't say "labor," though. Designing the robot was fun. I got to use my bright silver yarn, and I got to decorate it with a whole bunch of sequins and beads. I also thought it should be listening to some good music, because there's no point in being made out of circuits and wires unless you have your very own, custom stereo system.

So put on your favorite jams and enjoy crocheting this robot pattern that I wrote for you.

Materials

» **Worsted weight acrylic yarn** (exact brand/color used specified in parentheses):

   1 skein ruby (Caron Simply Soft: Rubine Red)

   1 skein hot pink (Caron Simply Soft Brites: Watermelon)

   1 skein silver (Bernat Satin: Sterling)

   1 skein black (Caron Simply Soft: Black)

» **Crochet hooks, sizes C (2.5mm), D (3.0mm), and E (3.5mm)** or size to obtain gauge

» **Polyester fiberfill** for stuffing

» **Black 15mm animal eyes (2)**

» **Black pearl cotton** for mouth

» **Sulky invisible machine sewing thread**

OPTIONAL EMBELLISHMENTS:

» **Flat, black sequins** for boots

» **Large, white flower sequins** for headphones

» **Small, pink flower sequins** for headphones

» **Pink Delica beads** for attaching all sequins

» **#8 silver seed beads** for eyebrows

1. Make the head.
Gauge: Head Rnd 1-3 with E hook = 3cm × 4.8cm at widest points.

Start with ruby and E hook.

Rnd 1: ch 6, 3 sc into bump of first ch from hook (a bit tricky — ch loosely and pull loop on hook a bit longer than usual), sc in bumps of next 4 chs, 3 sc in last ch, sc in next 4 ch on opposite side of beginning ch (should look just like sc because you worked into bumps on previous side). Do not join: 14 sc.
Rnd 2: (2 sc in next 3 sc, sc in next 4 sc) 2x: 20 sc.
Rnd 3: (2 sc in next sc, sc in next 4 sc) 4x: 24 sc.
Rnd 4-10: inc 3 sc evenly spaced: 45 sc.

Change to silver.
Rnd 11-13: inc 3 sc evenly spaced: 54 sc.
Rnd 14-15: work even in sc.

Change to ruby and D hook.
Rnd 16-19: dec 6 sc evenly spaced: 30 sc.

Time to give Robot a face. Examine the texture of the crochet. Notice ridges formed by rounds of crochet and valleys in between them. Look closer at the valleys and notice little posts running up and down from ridge to ridge. Put Robot's eyes in the third silver valley, 13 posts apart. Poke something pointier in there first to make the hole wide enough for the eyes to go through. Using a chain embroidery stitch, place Robot's mouth in the fourth silver valley, about 3 posts wide.

Rnd 20: dec 15 sc (decrease in every stitch): 15 sc. Stuff Robot's head very firmly. Body and legs won't have to be as firm.

2. Make the body.
Change to pink and E hook.

Rnd 21: inc 3 sc evenly spaced: 18 sc.
Rnd 22: work even.
Rnd 23-30: repeat (Rnd 21-22) 4x: 30 sc.
Rnd 31-32: inc 3 evenly: 36 sc.

Change to ruby.
Rnd 33: work even.
Rnd 34-37: dec 3 evenly: 24 sc.
Rnd 38: work even.

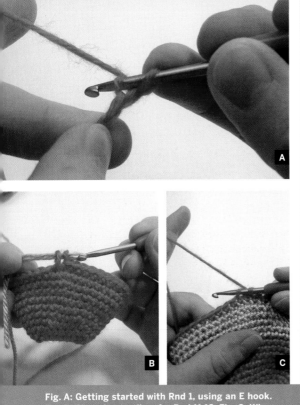

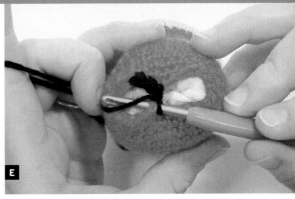

Fig. A: Getting started with Rnd 1, using an E hook.
Fig. B: Changing to silver for Rnd 11-13. Fig. C: When the pattern says "dec" (invisible decrease), first insert the hook under the front loop of the next step.

Fig. D: Robot's head after eyes and mouth are added.
Fig. E: The 12 stitches being skipped on the first leg and saved for the second leg. Fig. F: Rnd 1 of the second leg.
Fig G: Pulling up the loop to start the second leg.

3. Make the legs.

First leg: Change to black.
Rnd 39: sc in next 3 sc, sk 12 sc, sc in next 9 sc: 12 sc.
You'll make the second leg on skipped stitches.
Rnd 40-42: inc 2 sc evenly: 18 sc.

Change to silver.
Rnd 43: sc in next 8 sc, sc 2 hdc in next sc, hdc in next 4 sc, 2 hdc sc in next sc, sc in next 4 sc: 22 st.
Rnd 44: sc in next 8 st, sc 2 hdc in next sc, hdc in next 6 st, 2 hdc sc in next sc, sc in next 6 st: 26 st.
Fasten off. Weave in end.

Second leg: Change to black.
Rnd 1: Look at back of first leg, sk 4 sc of body and join black in next st by pulling up a loop. Sc in next 12 st: 12 sc.
Rnd 2-7: repeat Rnd 39-44 of first leg.
Fasten off. Weave in end.

Bottoms of feet (make 2): Change to silver and D hook.

Follow directions for crocheting into ch for top of head, but join Rnds as instructed.
Rnd 1: ch 4, 2 sc in first ch from hook, sc in next 2 ch, 3 hdc in last ch, sc in next 2 ch, sl st to join: 10 st.
Rnd 2: ch 1, sc in same space, sc in next 3 st, sc 2 hdc in next st, 2 hdc in next st, 2 hdc sc in next st, sc in next 3 st, sl st to join: 16 st.
Rnd 3: ch 1, sc in same space, sc in next 4 st, sc 2 hdc in next st, hdc in next 4 st, 2 hdc sc, sc in next 4 st, 2 sc in next st, sl st to join: 22 st.
Rnd 4: ch 1, sc in same space, sc in next 5 st, sc 2 hdc in next st, hdc in next 6 st, 2 hdc sc in next st, sc in next 7 st, sl st to join: 26 st.

Fasten off leaving 26" tail. Using tail, whipstitch bottom to each foot, lining up the 6 hdcs.

4. Make the arms.

Arms, starting with hands (make 2): Pink and D hook.

Rnd 1: ch 2, 6 sc in 2nd ch from hook: 6 sc.
Rnd 2: inc 6 sc: 12 sc.
Rnd 3: work even.
Rnd 4: sc in next 11 sc, dc 4 tog in next st: 12 st.
Rnd 5: sc in next 10 sc, dec: 11 sc.
Rnd 6: dec, sc in next 9 st: 10 sc.

Change to ruby.
Rnd 7: sc in next 8 sc, dec: 9 sc.
Rnd 8: sc in next 3 sc, dec, sc in next 4 sc: 8 sc.

Stuff hand firmly and continue to stuff arm as you go.
Rnd 9-15: work even in sc.

Change to black.
Rnd 16-18: work even in sc.
Fasten off leaving 12" tail. Pinch arm closed and whipstitch it shut. Weave in end.

Sew arms to the body, right under the neck with invisible thread.

5. Accessorize.

Headphones headband: Black and D hook.
Follow directions for crocheting into ch for top of head.
Row 1: ch 29, change to C hook, sc in 2nd ch from hook and in next 27 chs, ch 1, turn: 28 sc.
Row 2: sc in each sc: 28 sc.
Sew headband to head with invisible thread.

Ear muffs (make 2): Black and D hook.
Rnd 1: ch 2, 6 sc in 2nd ch from hook: 6 sc.
Rnd 2-4: inc 6 sc: 24 sc.
Rnd 5-6: work even in sc through back loop only.
Rnd 7-8: dec 6 sc evenly spaced: 12 sc.
Rnd 9: dec 3 sc evenly spaced: 9 sc.
Fasten off leaving 12" tail. Stuff lightly. Sew hole closed. With the invisible thread, sew muffs to the head at ends of headband with decrease sides facing head.

6. Finish.

You now have an adorably unadorned robot doll, which you can spiff up with as many sequins and beads as you like. Enjoy!

Beth Doherty has a bachelor of arts degree in Fine Arts from Columbia College in Chicago, where she still resides with her husband and cats.

The Comeback Craft

These two sparkly bangles make needlepoint cool again. BY KRISTINA PINTO

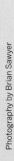

Photography by Brian Sawyer

Needlepoint has come a long way, mostly due to the introduction of more creative fibers and modern patterns. As one of the younger stitchers in any needlepoint class or shop, I've become interested in pushing the boundaries of this endangered craft. Encouraged by the knitting revival, I decided to develop hip and quick needlepoint projects to appeal to a more contemporary audience. If I have my way, needlepoint will be the new knitting.

These bracelets are easy, quick, and very chic. While stitchable in a weekend, they look much more complicated. Both pieces combine silk with metallic threads to create a look that goes with jeans as well as a black cocktail dress. Once stitched, take the bracelets to a needlepoint shop to finish for about $20 (or do it yourself).

Jade Wrist Cuff Materials

» **9" stretcher bars (2 pairs)**

» **9"×9" piece of eggshell 18-count mono canvas**

» **Kreinik metallic 087C medium #16 braid (10 meters)**

» **Needlepoint Inc. silk 872 (5 meters)**

» **DMC metallic 5270 (8 meters)**

» **Epoxy cement**

» **Thin hand material**

» **8"×3" piece of ultrasuede** in a coordinating color for backing (if doing the finishing yourself)

» **Laying tool** can be bought in a needlepoint store, or you can use a kebab skewer or unsharpened pencil for a fraction of the cost

» **Size 22 tapestry needle**

1. Assemble stretcher bars.

First, assemble your stretcher bars into a square frame and tack your canvas to it. You can cover the edges of the canvas with masking tape so the thread won't get caught and snag.

2. Stitch the cuff.

Once the canvas is framed, you'll start to stitch with 3-ply of the silk, using a length of about 18" in your needle. Beginning a third of the way up the canvas and about 1" in from the left, stitch the entire cuff in Giant Rice Stitch (Figure A). Use your laying tool to keep the stitches smooth and flat as you drop the needle in a hole, and work the bracelet in rows from left to right. Stitch each cross over 3 diagonal holes (Figure B) and make your rows long enough for the bracelet to go around your wrist in a cuff fashion. Your bracelet should have 7 rows of crosses (Figure D).

After completing the silk crosses, work the metallic green thread in diagonal stitches over the top of the silk, with each diagonal stitch crossing over 1 canvas hole (Figure B). No need to use the laying tool with this thread — just make sure it isn't twisted when you drop the needle in a canvas hole.

3. Finish the stitch.

Once you finish the Rice Stitches, backstitch vertically over 2 canvas threads at each end of the cuff to create a vibrant red line on both edges (Figure D). Use 3-ply of the DMC metallic for this step. Then, after taking the piece off the frame, fold the canvas back at each red end and whipstitch over a few canvas threads at the edges with 3-ply of the DMC to make the red thicker and to finish those edges. Whipstitch across each end of the cuff as many times as it takes for the red to cover the canvas underneath.

4. Finish the bracelet.

Now that the piece is stitched, you still have to make it into a bracelet. I recommend taking it to a needlepoint shop, where they can finish it for you by folding back the raw ends on the length of the bracelet and covering the back with a piece of ultrasuede. If you want to finish it yourself, I suggest following the instructions found at the DIY Network website (craftzine.com/go/cuff). When it's finished, find a can of tomato paste (or something of a similar diameter), wrap your cuff around the can, and put a rubber band around it to hold it in place for about a day. This will give the cuff shape so it doesn't fall off your wrist. Now it's finished!

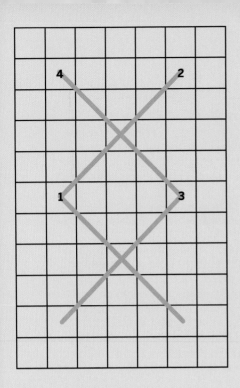

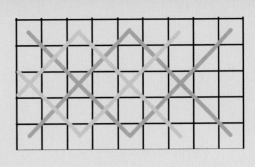

Fig. A: You will stitch the entire cuff in this Giant Rice Stitch. Fig. B: Stitch each cross over 3 diagonal holes. Fig. C: The basic bangle (described on the next page) is stitched entirely in this double-plaited cross-stitch. Fig. D: Your completed cuff bracelet should have 7 rows of crosses, as shown here.

Moonlight Bangle Materials

» 9" stretcher bars (2 pairs)

» 9"×9" piece of eggshell 18-count mono canvas

» Kreinik metallic 102 medium #16 braid (10 meters)

» At least 20 transparent or silver beads

» Splendor silk (1 card)

» 20-25 medium transparent beads

» 16-20 small transparent beads

» 1 large silver bead

» Laying tool

» Size 22 tapestry needle

» Beading needle

1. Frame up and begin.

This basic bangle is stitched entirely in double-plaited cross-stitch (see Figure C on previous page). After framing up the canvas in the same way as for the cuff, begin with a foundation of silk crosses (stitches 1-8 on Figure C) that are laid over 7 canvas threads, using 5-ply of the Splendor silk. Remember to use the laying tool for these stitches to keep the silk smooth and flat.

2. Stitch the double crosses.

Then, stitch the double crosses over the top (stitches 9-16) using the metallic braid. Weave the final stitch (15-16) under stitch 9-10. The entire bracelet is one row of these squares. Repeat blocks of double-plaited cross-stitch as many times as it takes to create a row that is long enough to go around your wrist without a clasp (it's a bangle bracelet).

3. Finish stitching.

Stitch a full block of double-plaited crosses in the silk, at both ends of the bracelet. Keep in mind that the stitch used for this bracelet will not fully cover the canvas, which adds some nice dimension to the piece. Coverage is assisted by the strategic placement of beads on the bracelet.

4. Add the beads.

Using your beading needle, tack down 1 medium transparent bead between each square, working across the bracelet, and tack down a border of the small beads, one at a time, around the silk squares at each end of the bracelet. Place 1 medium bead in the center of each of these silk squares. Now, take the canvas off the frame and cut the bracelet out, leaving about 4 rows of raw canvas around the perimeter. Fold the raw canvas back along both edges of the length of the piece and whipstitch them together. Then, fold back the short ends of the bracelet and stitch these down. Finally, attach the large bead between each end to create a bangle; you can stitch or glue this bead to the ends.

Kristina Pinto is a psychology instructor, fiber artist, runner, and mother. She has been stitching for 12 years. Her fiber projects can be viewed at: flickr.com/photos/threadgatherer/sets/433121.

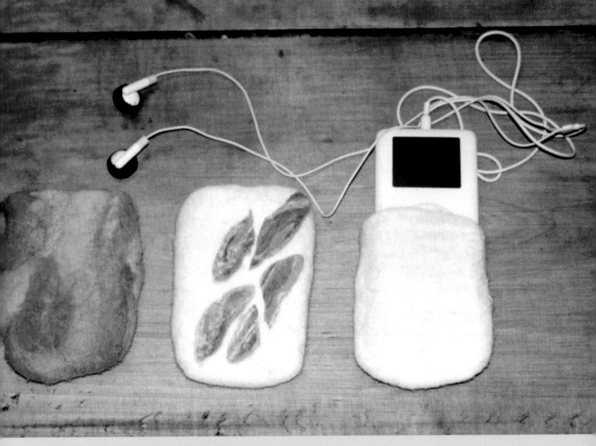

Rainbow Cocoon

Transform wool fleece into a felted iPod cozy.

BY EMILY DRURY

As a weaver and knitter, I love felting. Turning wool into felt is an exciting process in which the airy light fleece radically transforms into a dense material. Also, felting provides near-instant gratification compared to the pace of other fiber arts. This project will teach you the ancient art of felt making while showing you a quick, easy way to make a groovy iPod "cocoon."

  You can make something completely simple and classic, or get more complicated, adding texture and cool designs to your cocoon. Try using different-colored fleece or integrate pieces of wool yarn when wrapping the final layer of fleece (you'll find out how in Step 3). You can also try felting with different blends of fibers — for example, silk and wool — or sew or embroider decorations onto the finished dried cocoon.

Materials

» **1 oz. clean, combed, and carded wool fleece**

» **Thick, flexible flat plastic** (large yogurt top works well)

» **Scissors**

» **An old pair of tights or knee-highs, cut off about 10" from the toe**

» **2-3 quart bowl**

» **Hot water**

» **Ivory dish soap**

» **Textured surface that can get wet,** such as heavy duty bubble wrap (from pool cover, about 2 feet square), a piece of fiberglass screen, a washboard, a rattan placemat, etc.

1. Trace the pattern.

Trace your iPod onto plastic, rounding corners, then cut out "mold."

2. Cover the mold.

Place the plastic mold inside one of the stocking feet, and neatly fold over extra material.

3. Layer the fleece.

Pull out lengths of fleece, evenly fluff them out, and wrap around the covered mold, creating 3 layers of fleece going in alternating directions. For example, wrap the first layer horizontally around the mold, the second vertically, the third horizontally again. Evenly cover both sides and all edges. Carefully slip fleece-wrapped mold into second stocking, all the way to the toe. Tie a knot with excess material of the stocking (above mold), taking care not to alter rectangular shape of the mold.

4. Prepare to felt.

Fill the bowl ¾ full with hot (not boiling — you'll put your hands in it), very soapy water. Mix to make loads of bubbles.

5. Felt the fleece.

Dip stocking-covered fleece so that it gets soaked and soapy. Squeeze out excess water. Felt fleece by rubbing wet, soapy stocking on textured surface, using a fair amount of pressure, and working all sides while maintaining the integrity of the shape.

When it cools or dries out, repeat the above process. Continue to rub the fleece on the textured surface until it feels like it's hardening and turning into felt inside the stocking (15-20 minutes, depending on the surface and amount of pressure applied).

Untie and carefully remove cocoon — some felt might stick to the stocking. To modify the shape or surface of the cocoon, work it directly on the textured surface as described above. The longer you felt, the denser it becomes.

6. Finish.

Once you're happy with the shape and density of the cocoon, cut a slit in one of the short ends and remove the stocking-covered plastic mold. Continue to shape and work the felt further from the inside until you're happy with the look and feel.

Rinse soap out of the cocoon under running cold water, then carefully squeeze out excess water. Block to desired shape, and let dry (if necessary, stuff paper towels inside to help hold its shape).

Then have a blast decorating your cocoon!

Emily Drury is a textile artist living in Harrisville, N.H. She designs, produces, and sells a small line of handmade clothing and accessories.

Fiesta Explosion Flower Pots
Spiff up terracotta with vivid color and clay.

BY KATHY CANO MURILLO

Photography by Kathy Cano Murillo

F olk artists in Mexico are a shining example of what being crafty is all about. They can't simply run to Michaels or Jo-Ann Fabrics for a hunk of fuchsia clay. Instead, our south-of-the-border friends make the most of their resources. The results are shiny, happy, thrifty, and wickedly clever. From painted tequila bottles to glittered cigarette boxes to wood frames trimmed in bottle caps, the materials and concoctions are endless.

I channeled that infectious spirit for this project, and used only what I had on my art table to accessorize a bland flowerpot. Toothpicks and head pins are great for making microdots, and small paint bottles work well as templates for shapes. The secret to this look is to cram in as many contrasting primary colors as possible.

Materials

» **Large terracotta pot (1)**

» **Assorted acrylic paints**

» **Water-based brush-on varnish**

» **Cardstock**

» **Scissors**

» **Box of Sculpey polymer clay**
(or air-dry clay will work fine too)

» **Craft knife**

» **Cookie sheet and oven**

» **Head pin or toothpick**
(for painting small details)

» **Hot glue**

1. Prepare the pot for painting.

Base-coat your pot with a paint color of your choice, then varnish it and let dry. Set aside.

2. Make a template.

Use the bottom of the paint bottle to draw a silver-dollar-sized circle template on the cardstock. Cut out the template.

3. Make the circles.

Working on a clean, flat area, pinch off a hunk of clay, flatten it, and use the template to cut out a circle. Pinch off another piece of clay and roll it into a long skinny snake, about the size of a spaghetti noodle. Arrange it around the clay circle to look like a border, and cut excess with the craft knife. Add a swirly design of choice in the center and cut off excess. Make 10-12 more.

4. Bake the clay.

Bake in oven according to package directions — usually about 15 minutes at 175 degrees. Remove and let cool.

5. Paint the circles.

Base-coat the circles, then let dry. Add contrasting colors to the top ridges, let dry. Use the head pin to add dots and squiggles, let dry. Add a coat of brush-on varnish, let dry.

6. Attach the circles.

Hot glue the circles around the rim of the pot. Add more painted designs if desired, and a coat of high gloss varnish for that extra punch.

Variation: Add glitter to your circles, or make them in squares or other shapes. Use the clay to spell out words or other designs. Alternate the sizes of the circles for a crazier effect. If you absolutely must, use rubber or foam stamps on clay circles instead of swirlies.

Now set your festive pot on your windowsill, and you and your beloved plants will sing out *¡Que bonita!*

Other Crafty Chica Ideas

Want to whip up some more Mexicana crafts? Follow these simple tricks to spice up ordinary objects without breaking into your piggy bank. Visit your local import store to find these goodies — otherwise they can be found on eBay.

» Loteria — a popular Mexican bingo game — features cards with more than 50 images that can be cut up and glued onto boxes, tabletops, greeting cards, journals, and bottles.
» Get a hold of some old Mexican movie posters, reduce them by making color copies, then laminate them to make a set of placemats for your next Mexican meal.
» Use Spanish-language newspaper, gossip magazines, candy wrappers, and other paper items for scrapbook pages or collage art.
» Use milagros (small Mexican charms believed to make miracles come true) to make jewelry, or scatter them around a table to liven up a centerpiece.

Kathy Cano Murillo is the founder of craftychica.com and the author of *Crafty Chica's Art de la Soul: Glittery Ideas to Liven Up Your Life* (Rayo Books).

Making an Atomic Ball Clock

Put together a wooden rendition of this classic 1950s timekeeper. BY STEVE LODEFINK

The George Nelson ball clock is a neat little slice of mid-century art and architecture, but with the current licensed reproduction selling for around $265, I decided that if I were going to have one, it would have to be an unofficial version.

I made my first ball clock as an exercise in learning how to use a MIG welder, and ended up with a double-sized, welded-steel version of the clock. I was excited when asked to write this how-to piece, but realizing that most people probably don't have access to welding equipment, I decided to create a new ball clock that could be put together with all wood parts and assembled Tinkertoy style.

Conveniently, I found that I was able to gather up all the supplies for the clock with a single stop at a Rockler woodworking store, or a session on their website (rockler.com).

Materials

» **4½"×4½"×2" block of wood** for the central hub/body of the clock

» **36"×¼" hardwood dowels (4)**

» **2" hardwood balls (12)**

» **¾" shaft quartz clock movement**

» **Clock hands kit**

» **Epoxy cement**

» **Thin sheet metal or balsa wood** for oversized clock hands

» **Flat black spray paint**

» **220-grit and 400-grit sandpaper**

1. Shape the central hub.

Use a compass to mark a 4½" circle onto the wood block. Use a jigsaw or band saw to cut out the circle. You should make radial "relief" cuts before cutting out the circle (see Figure A) — this will prevent the blade from binding and burning the wood, and allow it to turn a tight corner. Next, use a belt or disc sander to remove any cutting marks, square up the hub, and make it smooth and round.

2. Drill the holes.

Drilling the holes for the radial spokes is the most critical part of this project. If the holes aren't all on the same plane and drilled directly toward the center of the hub, the clock will look catawampus.

First, drill a ¼" hole through the center of the clock. Next, scribe a line all the way around the outside of the hub, centered edge-to-edge. Using a protractor, make a mark every 30 degrees around the face, to make 12 evenly spaced marks (see Figure B). Transfer these marks down to intersect the line, and mark for drilling. Use a drill press to drill each radial hole 1½" deep, aligning the drill directly toward the center hole (see Figure C).

3. Prepare spokes and balls.

While you've got the ¼" drill bit in the chuck, drill a 1" deep hole in each of the 2" wooden balls. Typically, there will be a little flat spot on one side of each ball — this makes a great spot to drill the hole. Take care to drill straight into the center of the ball.

Next, use a fine-toothed saw to cut the ¼" hardwood dowels into twelve 12" lengths. Lightly sand the cut ends of each spoke.

If you prefer to stay truer to the original design, you can make the spokes from ¼" brass tubing, available at hardware and hobby supply stores.

4. Cut the clock movement recess.

Remove the locking nut and washers from the movement and insert the center shaft into the hole in your wooden clock hub. With a pencil, trace around the movement's case onto the hub. Draw the outline of the movement ¼" or so larger than it needs to be, to make fitting the movement easier. I used a router with a straight-cutting bit to excavate the recess for the clock movement, but you could also use a drill with a Forstner bit, and then square up the corners with a wood chisel. Cut the recess to a depth of 1¼", which will leave you with a ¾" floor. Test-fit the movement to make sure that there is enough shaft exposed to thread on the lock nut. Deepen the recess if necessary.

5. Make the hands.

To recreate the whimsical, oversized hands of the original Nelson clock, I started with a handset that was specifically made to work with the movement that I was using, and then augmented the stock hands with some oversized cutouts.

Balsa wood or thin sheet metal are ideal choices for a hand material, because they are thin, lightweight, and fairly rigid. The finished hands need to be light, or the somewhat anemic quartz movement won't be able to swing them around the clock dial. I cut the oval, triangle, and rectangle shapes for my new hands from a scrap piece of light-gauge

Fig. A: Making radial "relief" cuts before cutting out the circle prevents the blade from binding and burning the wood. Fig. B: Using a protractor, make a mark every 30 degrees around the face, to make 12 evenly spaced marks. Fig. C: Use a drill press to drill each radial hole 1½" deep, aligning the drill directly toward the center hole.

aluminum dryer duct that I had lying around the shop.

Mix up some epoxy, and cement the new hands in place over the old hands of the clock. Use something heavy to press the hands flat while the cement dries, and use waxed paper to keep the cement from sticking to your work surface. Once the glue dries, use an X-Acto knife to reopen the holes where the hands mount onto the center shaft of the movement.

Paint the hands with a few light coats of flat black spray paint. Flat black will help hide any imperfections in thin sheet-metal hands. If you make your hands from balsa wood, lightly sand the top surfaces with 400-grit sandpaper between coats of paint.

6. Finish and assemble.

Sand all wood parts with 220-grit, then 400-grit sandpaper. I made my clock body from a beautiful Mexican hardwood called bocote, which has a really nice natural color and figure, so I chose to finish the wood with a simple oil finish. I was happy with the color of the walnut spokes, so they received a coat of oil and wax too. If you plan to use a stain, varnish, or polyurethane topcoat on any of your wood components, apply it now. I gave the maple balls two coats of clear, non-yellowing acrylic lacquer.

Put a small amount of wood glue on the ends of the dowels before inserting them into the balls. Then press each ball/spoke pair into a hole in the hub, again using a small drop of wood glue.

Install the clock movement, securing it with its locknut. Install the hands, and thread on the little nut that holds them in place. Check that the hands don't interfere with each other as they move, bending them slightly to make adjustments if necessary.

I was able to mount my clock by simply hanging it on a nail, using the clock movement recess itself as a hanger, but if you want a more secure mounting, you could add one of those sawtooth picture hangers to the back of the clock. Or, drill a ¼" hole in the back of the clock hub to use as a mounting point.

Now you've got what is, in my opinion, better than an original Nelson clock. Not only does it have an accurate quartz movement that doesn't need a plug, but you get to fine-tune the size and finish to suit your needs.

Meanwhile, back at the ranch, you've got the perfect clock for that timeless spot.

Steve Lodefink makes web pages and knickknacks in Seattle.

Vintage Clock Conversion

Replace your clock's windup or wall-plug movement with a modern, accurate quartz unit. BY STEVE LODEFINK

The next time you're at a flea market, think twice before passing up that oversized gymnasium clock with a mysterious bundle of wires sticking out the back, or that cool $5 starburst clock just because it has a broken key-wind movement. The fact is you can easily convert these clocks to run with battery-powered quartz movements, which can be found at just about any craft or woodworking supplier, or salvaged from ugly clocks.

How you do this, and the necessary steps you must take, vary slightly from clock to clock, but the procedure is essentially the same.

Materials and Tools

» **Vintage clock**

» **Battery-powered quartz movement**

» **Epoxy cement**

» **Tweezers, small pliers, screwdriver**

1. Remove the old clockworks.

Typically, you must remove the clock's back covering along with its bezel or crystal assembly before you can remove the old movement. On my clock, the casing of the chassis on the back was fastened by two screws. After removing the chassis, I just had to pry up a few metal tabs to release the assembly.

Once the assembly is removed, you can take off the hands. Second hands are usually friction-fit and can be pulled off with tweezers or fingers; minute hands are normally retained by a small knurled nut; and hour hands are usually friction-fit to the outside of the shaft. Remove the nut and washer from the shaft.

Turn over the clock and remove the screws that hold the clockworks to the chassis or dial plate.

You should now be able to pull the movement out of the clock. Remove any unnecessary standoffs or mounting bosses that could interfere with the installation of the new movement. Cut and remove the power cord, if the clock has one.

2. Install the quartz movement.

Craft and hobby movements come with a variety of shaft lengths to accommodate different dial thicknesses. Get one that matches your clock dial; otherwise, the shaft may protrude too far, which could be a problem when it's time to refasten the crystal.

Insert the shaft of the new movement into the dial's center hole and install the washer and retaining nut.

3. Adapt the hands.

The design of a clock's hands is an important part of the clock's overall look and feel, and is something you may want to retain. Unfortunately, it's unlikely that the old hands will fit the new movement. To solve this, you essentially use the new hands to make a set of "adapters" for the old hands. How you do this depends on the shape of both sets of hands.

I adapted the hour hand by first enlarging the hole slightly to fit over the new shaft. I then cut the center hub from the new (donor) hour hand and epoxied it to the back of the original hour hand.

To adapt the minute hand, I enlarged the center hole to fit the flat-sided shaft of the quartz movement.

As with the hour hand, I adapted the second hand by cutting the center hub from the new hand and gluing it to the back of the vintage hand.

For certain designs, you may find that it makes more sense to cut a small, decorative element from the old hand and glue it to the new hand. Again, this depends on the shape of the hands.

Apply a coat of spray paint to the hands to hide the surgical scars, and install them on the shaft. Reassemble the clock and hang it wherever you wish — no extension cord required.

Steve Lodefink makes web pages and knickknacks in Seattle.

Raw But Refined

Homespun framing. BY MATT MARANIAN

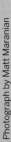
Some scrap plywood and a sheet of Plexiglas fitted with a few cents' worth of hardware is all it takes to get those must-frame items out of your drawers and onto your walls. This framing project can be built to suit, but for the sake of a starting point, I've sized the frame to accommodate an 8"×8" block print.

A note on materials: I like fir plywood because it's cheap and tends to bear the most exaggerated grain; ⅛" Plexiglas is available cut-to-size at a decent hardware store or can be purchased precut at most framing supply shops.

Using varnish on the wood will accentuate the grain and give it a satin finish. You may also leave the wood unfinished.

Materials and Tools

» One 16"×16" piece of ¾" A/C fir plywood

» **Fine-grit sandpaper** and **flat wood block**

» **Clean cloth**

» **Aqua ZAR water-based varnish** (optional)

» **Paintbrush** for (optional) varnish

» **8"×8" piece of ⅛" clear Plexiglas** (protective plastic sheeting left on)

» **Ruler**

» **Masking tape** and **Sharpie marker**

» **Electric drill** with $^{11}/_{64}$" bit

» Two $^{5}/_{32}$"×1" **eye bolts** with washers

» Two $^{5}/_{32}$"×1" **round-head machine screws** with hex nuts

» Two $^{5}/_{32}$" **acorn nuts**

» **Picture wire**

1. Prepare the surface.
Using fine-grit sandpaper wrapped over a flat wood block, sand the choice-side surface of the plywood smooth.

2. Apply varnish.
With a dry cloth, wipe the plywood completely to rid the dust created by sanding. Brush the surface with Aqua ZAR varnish. Allow the varnish to dry completely, lightly sand the varnished surface, wipe off the dust, and apply a second coat. Repeat. (Apply 3 coats of varnish in total.)

3. Mark the board corners.
Use a ruler to establish the centered placement of

the Plexiglas; each edge of the Plexiglas should be 2½" from each edge of the plywood. With masking tape, mark the corner points.

4. Mark the Plexiglas.
Using a Sharpie marker, draw 4 straight lines 1" from each edge of the Plexiglas, on the protective film (not on the Plexiglass itself).

5. Drill the holes.
Prop up the plywood a few inches above your work surface with scrap lumber or books. Hold the Plexiglas centered on the plywood using the taped corner guides, and drill a hole at each point where the lines drawn on the Plexiglas intersect. Drill straight through the plywood (taking care, of course, not to drill through the books).

6. Insert screws and eye bolts.
After drilling each hole, slip either a machine screw or an eye bolt into the shaft to anchor the placement of the Plexiglas while the remaining holes are drilled. Do not force the drill because the Plexiglas is likely to crack.

7. Place artwork.
Remove the tape from the plywood and remove the protective sheets from each side of the Plexiglas. Clean both the plywood and the Plexiglas clear of drilled shavings. Sandwich your artwork so that it's centered between the Plexiglas and the plywood.

8. Secure screws and eye bolts.
Slip the washers over the eye bolts and screw the hex nuts onto the machine screws. Slide the eye bolts and machine screws through the drilled holes from the backside of the frame, and secure them with the acorn nuts (the 2 eye bolts should be placed through the "top" holes in the frame to allow for picture wire and hanging). In order to fit the acorn nuts flush to the Plexiglas, the hex nuts may need to be adjusted.

9. Attach picture wire.
Run the picture wire through the eyes of the bolts, winding each end tightly to secure.

Matt Maranian is a bestselling writer, designer, and bon vivant whose books include *PAD: The Guide to Ultra-Living* and *PAD Parties: The Guide to Ultra-Entertaining* (both by Chronicle Books). He lives in New England.

Customized Kitsch
Create your own paint-by-numbers kit. BY LEAH KRAMER

In the 1950s, following the Depression and World War II rationing, Americans enjoyed leisure time and larger houses (meaning more square footage to decorate). And what did clever entrepreneurs do with these conditions? They mass-produced paint-by-numbers kits that allowed anyone to be an artist! But the kits didn't last long with hipsters, who became more interested in Andy Warhol's ideas on consumerism and mass production.

Fifty years later, Craftster member Annette Kesterson wanted to pay homage to this kitschy pastime. She took cheesy 70s pin-up images, used Photoshop to create black and white paint-by-numbers maps, printed them out, and included a paint brush and test tubes with the paint colors inside. She gave the kits as gifts to her friends, who were delighted. To create your own paint-by-numbers kit (PBN), snap a digital photo or scan an image and embark on the journey herein.

Materials and Tools

» **Computer, Adobe Photoshop, digital photo**

» **Piece of cardboard or mat board** as large as you'd like your PBN to be

» **3M Spray Mount glue**

» **Inexpensive acrylic craft paint** and thin paint brushes

» **Vials or small bottles with screw caps, labels** (up to 10; optional)

1. Make a backup copy.
Back up your image, then open it in Photoshop.

2. Save file.
Click File —> Save As. Save as a native PSD file.

3. Reduce number of colors.
Click Image —> Mode —> Indexed Color and set the dialog box settings to:
» Palette: Local (Adaptive)
» Forced: Black and White
» Transparency: unchecked
» Dither: None
» Preview box: checked
» In the Colors field, type in 6 and see if you think enough color has been preserved. If not, try 7, then 8, etc., until you're happy with the representation. Don't exceed 10 since this dictates the number of paint colors needed. Click OK.

4. Change color mode.
Click Image —> Mode —> RGB Color.

5. Smooth areas of color.
Click Filter —> Noise —> Median. Use the plus and minus buttons and the grabby hand to get a detailed part of the photo into the viewer at the top of this dialog box and set the settings to:
» Stroke Detail: 3
» Softness: 0
» Move Stroke Size slider to 1 and nudge higher until you get nice smooth areas of color. Click OK.

6. Repeat Steps 3 and 4.
So far, you've blocked off nice distinct areas, but if you zoom in really close, you'll notice that within those color sections it's a bit blotchy. So repeat Step 3 exactly as before to make sure that each section of color is solid throughout that section. Then repeat Step 4 as well to bring it back into RGB mode.

7. Duplicate layer.
Duplicate this layer to refer to later. Right-click on the only layer in the Layers window and click Duplicate Layer. Continue on this duplicated layer.

8. Place number labels.
Use the Text tool to put numbers in each region, corresponding with the paint colors. You should only need to use the numbers 1 through 10 for the 10 paint colors (or however many colors you chose in Step 3).

9. Create a B&W map of painting.
Use another Photoshop filter to create the black and white map of the painting. Click Filter —> Stylize —> Find Edges. This creates handy outlines of all the areas of color in the photo.

10. Adjust line color.
The outlines are drawn in colors. To make them a nice consistent light gray, do the following:
Click Image —> Adjustments —> Desaturate.
Click Image —> Adjustments —> Brightness/Contrast and move the Contrast slider to 100.
Click Image —> Adjustments —> Brightness/Contrast and move the Brightness slider to 100.

11. Print and mount image.
Print the PBN map out as large as desired and affix it to a stiff piece of cardboard or mat board using 3M Spray Mount glue.
   To turn this into a kit, put the 10 colors of paint into labeled vials or small bottles. Toss in a few thin paintbrushes, and you've created a truly one-of-a-kind objet d'art waiting to happen!

Leah Kramer thinks she's inhaled a little too much glue over the years because she's become inordinately attracted to crafts that are clever, ironic, irreverent, and offbeat. She's the founder of craftster.org and author of the book *The Craftster Guide to Nifty, Thrifty, and Kitschy Crafts*.

Cider Rules

Make and bottle your own hard cider.

BY HOLLY GATES

Photography by Holly Gates

W hat's more American than warm apple pie? Turns out to be a big jug of hard cider. Apple trees introduced by English colonists grew especially well in the Northeast, and were spread across the country by farmers and Johnny Appleseed. Before refrigeration, fermenting juice from apples was the easiest way to retain most of their useful calories. The resulting alcoholic brew kept much longer and, much like its malty cousin beer, served the crucial need for safe hydrating beverages in the ages before advanced water treatment.

Almost all hard cider on the market today bears little relation to the complex nectar you can make at home. After you taste the (fermented) fruits of your labor, commercial cider will seem like weird apple candy dissolved in lighter fluid. After adjusting to the subtle and intriguing new flavors, chances are you'll never go back.

Materials and Tools

- » Apples
- » Sulphite tablets
- » Yeast
- » Bottles
- » Fermentation vessel
- » Sanitizer, like iodophor
- » Fruit press
- » Airlock
- » Apple shredder
- » Carbonation equipment

1. Gather fruit.

The best cider is made from a blend of apples. Look at a long list of apple types, and try to find some varieties that are listed as good for cider. Crab apples are a welcome addition to the mix, as they will inject some wildness into the blend. Try to get 4-10 varieties of apples in your blend, with a range of flavors.

Specialty cider apples have flavor descriptors like "bitter-sharp," which describes an apple high in tannins and acid. These apples taste pretty bizarre if eaten out of hand, sometimes almost medicinal,

and have interesting names like Foxwhelp and Kingston Black.

If practical, the clear winner for selection and freshness is a self-pick orchard. Another good option is a farmers market, with store-bought apples a distant third choice.

Yields will vary, but we got around 1L of juice per 4½ lbs. of apples. Homebrew equipment and supplies are mostly based around 20L batches, making that a convenient number to shoot for.

Let the apples rest for a week or two before being milled. This is called mellowing, and will increase yield.

2. Shred.

First, wash your apples off with a hose. No sense in concentrating pesticides and bird crap into your cider.

You will need some way to shred, pulverize, or otherwise rend your apples. If your grandparents happen to have an antique cider press with shredder rotting in a barn in Maine, all the better. Otherwise, a $30 garbage disposal from Home Depot mounted on a piece of plywood and discharging into a bucket is pretty much the perfect thing to pulp apples. A food processor will certainly work, but will take forever.

The output of this step is an apple mush called pomace.

3. Press.

As your pomace comes out of the shredder, put it into your press. On the slatted cylinder-style press, the shredder can shoot apple gunk straight into the muslin-lined press bucket.

The other main family of press designs consists of a rectangular frame with the pomace loaded in horizontal, cloth-wrapped packs called cheeses, interleaved with wooden grates, and pressed with some type of ram (usually hydraulic). You can probably knock something together along these lines with a few 2x6s, some lag screws, and a car jack.

Transfer collected juice to your fermentation vessel.

4. Ferment.

The most traditional cider fermentation utilizes the wild yeasts naturally present on apple skins. To go full-on rustic, slosh the juice into an old wooden bucket, throw a lid on it, and come back in the spring.

For more predictable results, kill wild bugs in the

Fig. A and Fig. B: Gates and his friends spend a day at Red Apple Farms in Phillipston, Mass., where they pick 160 pounds of fruit. Fig. C: An old cider press/shredder found in Grandma's barn in Maine is used to shred the apples. Fig. D: The press in action. Fig. E: A bucket of pomace (makes good compost) and freshly pressed juice. Fig. F: Pouring juice into a fermentation vessel. Fig. G: A vial of White Labs liquid yeast.

Fig. A: Fizzy cider is siphoned into bottles from a soda keg with a CO₂ cylinder and regulator. Fig. B: Using an antique letterpress to design labels gives your bottled cider an authentic and appealing look.

Fig. C: Beautifully designed labels make home-bottled cider as enjoyable to look at as it is to drink (well, almost). Nicely packaged cider also makes an excellent gift.

juice, then introduce your own strain of yeast.

For a vessel, use a glass carboy or 5-gallon bucket with a hole drilled in the lid for the airlock. Dump the juice into your vessel, then dose with 5 sulphite tablets per 20L, and put the airlock in.

Twenty-four hours later, take off the airlock and pitch yeast. Ale, mead, and cider yeasts leave more sweetness in the cider, while champagne or wine yeast eat up more sugar and leave the product drier.

Use standard homebrew methods for keeping extraneous bacteria and yeast out of the product from this point on. Sanitize everything the juice touches using iodophor and a fermentation airlock to exclude airborne microorganisms.

Keep your vessel within temperature range for the selected yeast during primary fermentation, which'll take 1-2 weeks with ale yeast. Afterwards, you can optionally siphon the fermented liquid to a second, sanitized, airlocked vessel and leave for another few weeks; this'll produce a clearer, more mature cider.

5. Bottle.

For still cider, just siphon into bottles and cap. For effervescence, force-carbonate in a soda keg with a CO₂ cylinder and regulator before bottling. If you lack access to force carb equipment, dissolve ¾-1 cup of corn sugar per 20L batch in a few cups of boiling water, cool, then thoroughly mix into the cider before bottling. Sanitize bottles and equipment.

6. Label.

Nice labels make your cider look as good as it tastes. I prefer antique letterpress equipment, but you can also inkjet labels or order them on the web.

Further Frontiers

Apparently, cider quality significantly improves with aging, especially if done in oak barrels. Personally, none of the cider I have made has survived long enough to tell. Serve chilled and enjoy!

Resources:
positron.org/brewery
breworganic.com
Cider: Making, Using & Enjoying Sweet & Hard Cider by Annie Proulx & Lew Nichols (Storey)

Holly Gates is a veteran of MIT and its infamous Media Lab, and is currently a hardware engineer at E Ink trying to make electronic paper a reality.

Lanterns-a-Go-Go

Convert electric Chinatown party lights to battery power. BY MISTER JALOPY

Photography by Mister Jalopy

Swanky picnics, exotic tailgate parties, or cherry blossom backyard springtime wonderlands all benefit from a set of Chinatown lanterns — but what if there's not an electrical outlet within extension cord reach? What if you want your pup tent, phone booth, or shopping cart to look like an opium den? Convert to battery power!

RadioShack sells a battery for a remote control car and a connector repair kit that are perfect to convert a string of lanterns to be 100% mobile. Rather than soldering the whole mess together, use butt splice connectors that crimp the wires together with a single tool — a combo wire cutter, stripper, and crimper.

Will you install your lanterns as 70s-style lowrider headliner pom-poms? I hope so. I did.

Materials

» **String of 16 pagoda lanterns**
A Chinatown staple. Quality control
varies, so test before you leave the store.

» **RadioShack R/C Car Battery Pack
(#23-322)** and **Connector Repair Kit
(#23-444)** Conveniently, RadioShack
sells the perfect interface between the
battery pack and the lanterns.

» **Combo tool** (wire strippers, crimper,
and cutter)

» **RadioShack connecting wire
(#278-567)** OK, the black and red
18-gauge speaker wire is more expensive than thinner, clear, insulation wire
that would accomplish the same task.
Greater compromises have been made
in the name of aesthetics.

» **Butt splice crimp connectors**

1. Cut wire to length.

To determine the wire length between lanterns,
decide the total length of your new string and
divide by the number of pagoda lamps minus
one. For example, I wanted a 252" (21') string of
15 lights, so my formula is 252/(15-1)=18" for
each length. The string of lights comes with 16
lanterns, so why did I pick 15? In case of mishap,
I will have a backup lantern.

2. Separate and strip.

For each length of wire, split the insulation with your
fingernail to separate the red and black wires. Use
the strippers to remove ⅜" of insulation from the
wire segments and ⅝" from the lantern pigtail wires
(Figure C).

3. Last lantern is connected first.

After stripping insulation, twist copper wires to
tighten, then slide on the butt splice connector.
Use the crimp dies on your do-all tool to smash the
connector to your pre-cut, pre-stripped wire. (The
lantern wire is sooo fine that you need to fold it back
on itself and twist to bulk it up so there is something
to crimp to.) Put lantern wires into the other side of
the butt connector and crimp to connect (Figure E).

Is there a more elegant, more labor-intensive,
longer-lasting solution like soldering and using
shrink-wrap? Of course! Isn't there always a more
artful method that requires greater dexterity, more
experience, and additional tools?

4. Connect remaining lanterns.

Twist one lantern wire to red and the other to black
(Figure D). Then crimp a butt splice connector to
each wire pair (Figure F). If your plan was to build
a light string with only 2 lanterns, you are done! Otherwise, keep clipping, stripping, and crimping like
an 80s hair salon. Enlist helpers to construct 14 of
these wire segment/lantern pieces.

5. String lanterns together.

The lanterns finished in the previous step are daisy
chained one after another until the light string is
complete (Figure B).

6. Ding-Dong-Done!

Crimp the RadioShack Connector Repair Kit to
the end of your lantern string and you are ready to
connect the battery. Cross your fingers! Did all the
lanterns light? No big deal, I had a few duds, too.

Quick Tip: Troubleshooting
Disconnect battery before attempting repairs.

If none light:
» Check that Connector Repair Kit connectors are tight to
the lantern string.
» Charge battery.

If some don't light:
» Check the bulb.
» For any errant lanterns, cut off the butt splice
connectors, re-strip wires, re-crimp.

If lantern blinks furiously:
» Sounds like a blinker bulb — replace with a non-blinker.

**Mister Jalopy breaks the unbroken, repairs the irreparable,
and explores the mechanical world at** hooptyrides.com.

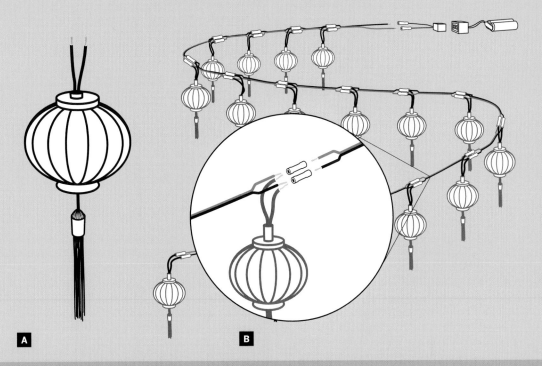

A

B

Fig. A: Cut all 16 lanterns from the string, leaving two pigtail wires like a double-fused cherry bomb. Fig. B: Crimp as you go to connect all 15 lights in parallel. Fig. C: Cut 15 wire segments to length and strip insulation to reveal ⅜" of copper wire. Fig. D: Crimp butt connectors to a length of wire, then to last lantern. Fig. E: To make more segments, twist one lantern wire to red and one to black, then crimp. Fig. F: Lantern segments are daisy chained like so.

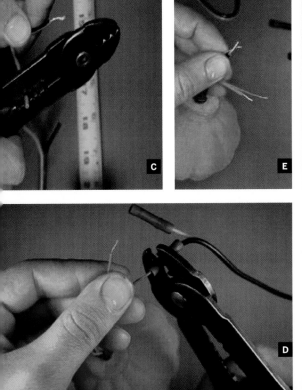

C

E

D

F

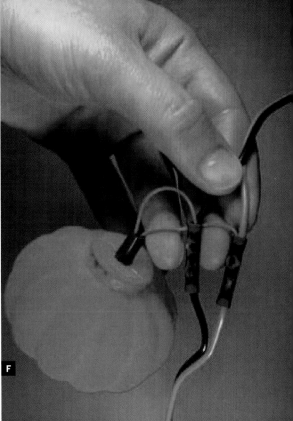

Random Screen

Turn your recycling into art. BY ARAM BARTHOLL

The Chaos Computer Club's *Blinkenlights* project in 2001 — which turned an office building in Berlin's Alexanderplatz into the world's biggest interactive computer display — inspired me to think and work on low-tech analog and mechanical screens.

Besides being able to play Pong using your cellphone on an 8-story-high, 18-window-wide display, it was possible to send small, self-made movies called "love letters." This was the way my brother asked his girlfriend to marry him.

Since then, I've wondered how to play low-res pixel movies on a small scale without any standard screen technology, and built *Paper Pixels*, a punchcard-controlled mechanical analog on an 8x8-pixel screen. To push things a bit further, I started thinking about candles and pixels, and the concept of *Random Screen* popped into my head.

Random Screen is a non-controllable, 4x4-pixel screen run by tea candles. Each pixel is a 5"×5" box made of cardboard, which is open at the back and closed with translucent film as a projection screen at the front. A modified beer or soda can is transformed into a kind of vent and driven by the rising heat from a tea candle, which also serves as a light source. A window is cut into the beer can, which casts the candlelight while turning at its individual frequency, like a lighthouse lantern. The

brighter and bigger the candle flame, the faster the can turns to switch the designated pixel on and off. The light of each pixel fades smoothly in and out.

1. Get ready.

First of all, have a good time and drink the beer (or soda). If you're planning to build several *Random Screen* pixels, it might be wise to invite some friends so that you don't get too wasted while preparing the materials. I used to store some beer cans in our shared office fridge, which is a very easy and quick way to get them emptied.

2. Make the pinwheel and stand.

Cut off the top and bottom of the beer can and shorten it to 5". For 9 vent rotors, cut from the top in equal distances 1¾" into the can, at an angle of 12 degrees, with shears or sharp scissors (see Figure A).

Punch a hole through each rotor near its top end, so that the tops of all rotors can be drawn together and a screw can pass up through the holes to hold them together. This part needs some patience, and be careful. Lock the screw in place with a nut.

Then make a simple wire stand (see Figure A). The base should fit around the tea candle, and the other end should be bent in and up so that the can will hang over the middle of the candle. Attach a needle to the end with the cable connector, checking to be sure that the pinwheel can spin easily.

3. Cut window.

Cut a 2"×2" window in this modified can and cut in some zigzags to make the light fade in and out smoothly. Run a test and place a tea candle inside to see if the can hangs straight. Make sure that the can is able to turn freely. You might have to work and bend the material a little bit.

4. Build the pixel boxes.

Cut and/or fold the cardboard to build a 5"×5" pixel box 7" deep. In order not to cast shadows or any movement onto the front pixel screen, make a middle wall inside the box to separate the back candle space 4" from the front projection space, leaving 3".

Materials

» **.5L or 16 oz. beer/soda cans (16)**

» **1mm cardboard** or (even better) some similar fireproof material

» **Translucent paper, film, foil, or even glass** I used inkjet backlight print film.

» **Stiff wire, a needle, a cable connector, a Phillips machine screw and a nut**

Photography by Aram Bartholl

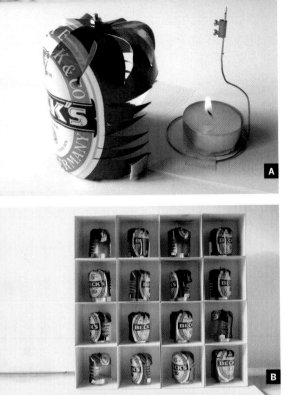

Fig. A: The completed beer can pinwheel and stand. The notches in the screw head sit nicely on the top of the needle attached to the wire holder. The needle reduces the friction on the pinwheel so it can spin faster.

Fig. B: You can make a single pixel, or build a bunch of them for stacking. Fig. C: The completed project makes for a gorgeous, constantly changing light display.

A 2"×2" window covered with the translucent film diffuses the passing light from the candle onto the main front projection screen.

Place the stand plus modified beer can in the open back of the box. Light the candle to see if everything works. If you built more than one pixel, you can just stack them on top of each other. (Of course it is also possible to build a 9- or 16-pixel *Random Screen* in one piece, but I like to make each pixel separately as a module you can play with.)

5. Turn out the lights!

A dark room is needed to obtain satisfying results. Light all the candles and watch your work of art flicker in the dark.

WARNING: A 16-pixel *Random Screen* produces quite a bit of heat, so be careful with flammable materials and never leave it unattended. Don't burn your house down!

Resources: datenform.de/rscreeneng.html
Read a profile about the author in *MAKE, Volume 07,* page 23.

Quick Tip: Cutting and Tweaking

Watch your fingers! Aluminum cans have a very sharp edge when cut. Be particularly careful when gathering the rotors together. If your can doesn't hang straight, add some of the bits you cut out as counterweights. Feel free to move the materials around to make for the best fit and to maximize balance and spin.

In his art projects, Aram Bartholl tries to transform objects and behaviors from the digital era back to the analog, mechanical, and physical world.

Bob Parks
Recycle It

>> Bob Parks is the author of *Makers: All Kinds of People Making Amazing Things In Garages, Basements, and Backyards.* While renovating his home in Brattleboro, Vt., he has learned that many sins can be hidden with good room lighting. bobparks@yahoo.com

Switched-On Cookware

Homemade lamps are the gateway drug to more sophisticated — and more dangerous — dalliances with electricity. Those folks throwing long sparks from Tesla coils and Van der Graph machines? They all started with dorky little projects involving 25-watt tulip bulbs. The wiring is easy to learn and requires only a few safety hints (more on that later). You can put a light bulb into anything, particularly into things found around the kitchen. Nothing's better than a mysterious glow emanating from a dead appliance or iconic soda bottle.

Amateur lampmaker Eric Gillin, for instance, created superior mood lighting for his apartment from a 1940s Cuisinart mixer. "It packs a huge surprise for visitors," says the Manhattan-based magazine editor. He hollowed out the flea-market find, installed two small lamps where the beaters should be, and added a translucent red glass bowl to diffuse the light. "People get a good laugh. It's fun to pervert something to have an entirely different purpose."

Kitchenware works well as the raw material for improvised lamps because it is instantly recognizable. Crafters have made lights from liquor bottles, chandeliers from champagne glasses, and luminaries from catsup jugs. Photographer and tinkerer Warren Armstrong made 1,000 points of ambience by putting a bulb into a cheese grater. Mike Knapp used Mountain Dew cans as fixtures for an attractive track lighting scheme. Argentine designers Adrian Lebendiker and Ricardo Blanco created an overhead light by meticulously cutting up a plastic soda bottle into thin strips and using the fuzzy material over a bulb to cast a soft greenish glow.

Phoebe Palmer, the high priestess of enlightened cookware, has made hip lamps from cereal bowls, thermos stoppers, casserole dishes, ice cream scoops, gravy boats, fondue pots, and a particularly fetching standup number made from 27 jello molds bolted end-to-end. "It's just one of the deep-seated female preoccupations," says Palmer, an artist based in San Luis Obispo, Calif. "I started out doing concrete work using jello molds. That struck me as funny. Then I started specializing with the lamps. I guess I'm drawn to the 50s kitchenware that was around when I was coming up." She often sells her works for $100 and more at haute houseware stores along the West Coast.

Some of Palmer's pieces skirt the line between Salvador Dali and Salvation Army. Let's be honest, we've all seen ugly novelty lamps. Wagon wheels, rustic milk jugs, and carriage oil lamps were made into willfully hideous living room appointments in the 1970s. I won't even mention previous eras (like the dad's high-heeled leg lamp in Jean Shepherd's *A Christmas Story*). But right now, kitchen stuff presents safe aesthetic ground because it is kitschy in a *cute* way. That the postwar kitchen was such a politically loaded place is something that Palmer can parody with her hacked-up atomic-era junk.

And for most of us, it's the easiest route to cool and unique lighting options. Gillin created his Cuisinart mixer after coming up empty at the store.

"Everything that was a little different was too expensive," says Gillin. "And everything cheap had no style, no soul." With an idea to transform an old mixer, Gillin walked down to the Strand bookstore and bought the Time-Life book on wiring. He soon made four or five lamps out of old blenders for friends, and moved on to a toaster with fluorescent tubes. It had an eerie glow — sort of like the car in *Repo Man* — but didn't generate enough light to be useful.

The essential tools for lampmaking are small pliers for cutting and stripping wire, a flat-head screwdriver, and electrical tape. Any hardware store will sell lamp sockets for under a dollar. It's always

Right: Phoebe Palmer reincarnates kitschy old kitchen treasures into fabulous new lamps.

Photograph courtesy of Phoebe Palmer

LID FOR COCKTAIL FORK HOLDER

ASHTRAY

PUNCH BOWL

SUGAR BOWL

CANDLE HOLDER

FAUX IVORY COCKTAIL FORK HOLDER

CEREAL BOWL

PEACH ENAMELED SERVING BOWL

COPPER-PLATED COLANDER

best to start with a new electrical plug and cord set than to use a ratty old cord from the original appliance. For a few dollars at Lowe's or Home Depot, you can even get a socket, switch, and cord combo (no electrical work at all!).

On one of his blender projects, Gillin took the design to the next level by mounting a switch behind the actual buttons for the blender. When delighted visitors pushed "Mix" or "Chop," they activated a light in the glass pitcher. "Dealing with electricity really isn't that hard," he says.

Well, yes and no, counters electrician Howard Oven of Belmont, Mass. Nonprofessionals often commit a big transgression when wiring their own lights — they connect the hot wire to the wrong terminal on the light socket. It's called reverse polarity, and regarding light fixtures, it means "the outside of the fixture — the screwshell — will be live," he says. "It could have devastating effects." A miswired lamp could shock someone who screws in a new bulb and touches the screwshell. And if the metal socket accidentally touches the metal housing of your blender, the whole device could become live.

The best way to prevent this, says Oven, is to buy a continuity tester (about $15 at RadioShack) and make sure the hot wire connects to the inside contact at the base of the bulb. The neutral wire — the one with a ribbed pattern in the insulation and a fatter conductor down on the plug — hooks to the outside. When done, double-check all your connections with your nifty new tester.

The other safety consideration is heat. Traditional incandescent bulbs will generate enough heat to make a fire if placed in contact with plastic, wood, or paper. Gillin always uses blenders with glass pitchers for this reason, making his yard sale hunts a little tougher. Palmer advises making sure there's a way to vent the hot air from building up inside ad-hoc lamps. She typically drills a 1" hole in a metal bowl when using it as a lampshade.

On a practical level, the heat from bulbs will melt most adhesives or hot glue used to fix the socket in place, creating a crooked setting for your bulb after a few hours. Personally, I once had a dozen hot-glued gumballs fall off a lampshade after firing up a small-wattage incandescent. Gillin, who grew up working in his mom's catering company, says he once melted a number of sheets of green gels trying to make his blenders look like they were full of margaritas.

To avoid the heat issue, consider a different type of

This DIY lamp was made from an old washing machine barrel by Dominik Enenek Zacharski.

bulb. A good option would be to use a compact fluorescent lamp (CFL), which creates less heat, uses a quarter of the electricity for the same light, and comes in both small and medium screwbases. High-quality CFLs can be spendy ($5–$7), but their manufacturers mix three or more phosphors for a soft glow that looks as good as incandescent light. Advanced DIYers have also begun using LEDs, which are 12 times as efficient as filament bulbs and run very cool. The strange thing is that no affordable 120-volt AC LED fixtures are currently on the market, but hacking your own is fairly easy and safe. It involves finding a surplus power supply for low DC current, soldering a few resistors, and repurposing some LEDs from a dollar store (to see how some brave lamp trailblazers did it, check out craftzine.com/go/ledbulb).

Whatever bulbs or technology you use, the best part of making a lamp is the satisfaction of being your own product designer. Lighting up your own kitchen doo-dad or appliance for the first time is like the punchline to a joke. If it's wired right — and it will be if you take minimal care — you'll see a nice warm glow and get a chuckle every time you enter the room. ✂

Photograph by Dominik Enenek Zacharski/purr.org

Clothing
reckon $19

Art
menta $15

Toys
sweetestpea $20

Jewelry
DownToTheWire $36

Toys
Curster $18

Bags And Purses
dejavu $25

Accessories
mogo $20

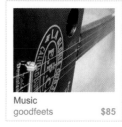

Music
goodfeets $85

Glass
ifusesolutions $100

Housewares
littleoddforest $8

Ceramics And Pottery
lusterbunny $30

Art
Magicjelly $15

Bags And Purses
macialaw $15

Children
patriciatomaszek $50.50

Jewelry
preciouspups $24

Ceramics And Pottery
tashamck $38

Etsy seller: Craftsman.etsy.com

"If I have learned one thing, it is that when you sell handmade stuff you're selling the hand that made the stuff. What I mean is that the customer is buying something they like from someone they like."

The items above were selected from the over **200,000 handmade goods for sale** on Etsy. Find any of the items above for sale, each in their own shops. Go to: shopname.**etsy**.com

Find the **unique**. Support the **independent artist**.

Accessories
Saltygal $15

Bath And Beauty
spazzspun $6.50

Clothing
stopsandstarts $26

Accessories
tinymeat $12

Children
1boy1girl $36

Clothing
tiinateaspoon $100

Art
ashleyg $20

Accessories
bonspielcreations $13

Jewelry
plutostar $60

Paper
rainyprints $12

Supplies (handmade)
pancakeandlulu $16

Housewares
supastarr $10

Toys
vladmaster $18

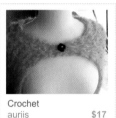

Crochet
auriis $17

Furniture
mclfurniture $475

Plants And Edibles
Yummy $15

Etsy seller: Dyno.etsy.com

"A graphic designer by profession, I prefer the all-encompassing 'I make things pretty' to describe what I do. While I respect graphic design, clothing design has always been my driving, visceral passion."

Make things? Join the community of over 22,000 individual artists, artisans, crafters, and makers of all things handmade who are selling their goods on Etsy. It doesn't matter what you make, as long as you do it yourself.

Your place to buy and sell all things handmade. etsy.com

After fashion designer Winter Rosebudd became a mom, she realized that cool toys for tots were hard to come by. So when she inherited a scuffed-up Lil Tyke kiddie car from her neighbor, she transformed it into "something fun that matched our home and punk lifestlye." Although it looks extraordinary, Rosebudd didn't need much to make it — just some funky fabrics and trims, Krylon Fusion spray paint, and lots of hot glue.

On punkymoms.com, she now organizes group craft-alongs for parents who want to revamp their plastic eyesores together. For details on how she created her groovy gothmobile, visit punkymomsforum.com/smf/.

Photograph by Susan Sheridan